AF665504

Influences of
the Forces of the Zodiac
on the Cultural Development
of Mankind

Cover picture:
Albrecht Dürer: Imagines coeli septentrionales cum duodecim imaginibus zodiaci
(Northern celestial images with twelve zodiacal images)

Roland Schrapp

Influences of the Forces of the Zodiac on the Cultural Development of Mankind

FSC
www.fsc.org
MIX
Papier aus verantwortungsvollen Quellen
Paper from responsible sources
FSC® C105338

Bibliographical information of the Deutsche Nationalbibliothek (German National Library): The Deutsche Nationalbibliothek lists this publication in the Deutsche Nationalbibliografie (German National Bibliography); detailed bibliographic data are available on the Internet at http://www.dnb.de.

Production and Publishing: BoD – Books on Demand, Norderstedt (Germany)

ISBN: 9783755717072

Contents

Zodiacal Images – Fantasy or True Vision?

When we turn our gaze to the sky on a cloudless night, we are confronted with a truly amazing splendour of countless glittering and twinkling stars in the firmament. This sight can touch us deeply inside, and even the most rational, sober minds will be amazed at such moments. Mostly we then try to find one or another group of stars known to us, like the "Big Dipper". Those who have retained some astronomical knowledge from their school days will probably also look for the neighbouring "Little Dipper", whose "handle" begins at Polaris, which is located at the northern celestial pole and around which, as a result of the Earth's rotation, the entire starry sky revolves in the course of a day. Some will even locate one or more of the twelve signs of the zodiac, and thus be able to trace the course of the sun's path in the sky, the so-called ecliptic.

From time immemorial, mankind has been particularly interested in the constellations lined up along the ecliptic. The images of the zodiac are among the oldest representations of the starry sky that have been handed down to us. But how did they come into being? To the enlightened modern man it seems plausible that the people of earlier times simply combined groups of individual stars by straight connecting lines to those abstract figures, which we still find today in our star charts, and that these figures served afterwards only as an incentive to decorate the starry sky with more natural pictures imaginatively. But did it really happen that way? Did the early astronomers first create abstract line figures and only afterwards add the vivid, more naturalistic images of Aries, Taurus, Gemini, etc.?

Would it not be possible that everything happened exactly the other way round? Couldn't the zodiacal images even be much older than the abstract figures created by connecting lines, so that the latter descend from the images and not the images from the figures?

Were the people of earlier times perhaps much more different in their soul life than we imagine today? Our natural scientific study of man, anthropology, examines the evolution of man mainly in terms of his physical development. Little consideration is given to the fact that quite substantial changes of the soul life and the perceptive faculties went along with it. We consider the people of earlier times to be much more fanciful than we ourselves are, and the fact that all ancient peoples were convinced of the existence of higher spiritual beings, whom they called gods, appears to us as a proof of how much the mankind of antiquity and even earlier times tended to an almost childlike fantasy.

However, this is opposed by a fact that cannot be explained away: It is precisely those people who spoke of images in the sky and of gods who created buildings such as Stonehenge, the Egyptian pyramids and the architecturally perfect Greek temples, about which we are deeply astonished today. Our scientists are still occupied with the question of how such gigantic and geometrically exact structures could have been created with the simple means available at that time. There is one thing these people certainly could not have been: dreamers out of touch with reality.

Shouldn't we rather assume that people at that time experienced their external world differently than we do today? Did they perhaps face the world and its phenomena not only with their physical sense organs and an abstract thinking, but with their full and whole soul, with a much deeper sensitivity and receptivity, which enabled them perceptions that have faded for us today? Did their inner perceptions perhaps extend even beyond those sensory limits which so strictly enclose our consciousness today? Did people possess a kind of "clairvoyance"? After all, even in the late Middle Ages we find reports about individual persons who are said to have possessed "the second sight". So, were the star images really seen and not simply constructed mentally? Rudolf Steiner gave the following information:

"The one who would simply draw what he sees there in the vastness of the world, would get nothing. A mere drawing of the starry sky, as it is done by today's astronomers, leads to nothing. But if one confronts the whole human being with a full understanding of the cosmos with this cosmos, then images are formed in the interior of the soul concerning these star clusters, as one sees them on old maps, when still from the ancient, instinctive clairvoyance the Imaginations were formed, and then one gets an Imagination[1] *of the whole cosmos. [...]*

People today have basically no idea of how one once saw into the universe in ancient times, when people still had an instinctive clairvoyance. Today it is considered that the different drawings, the pictures, the Imaginations, which have been made of the zodiacal images, have arisen from fantasy. They are not. They were felt, they were seen by facing the cosmos. The progress of mankind demanded that this instinctive, this vivid, this imaginative view was dimmed, that it was replaced by the intellectual view, which liberates man, out of which, however, if we want to be whole human beings, such a view of the universe must again be attained, which advances to Imagination, but now with full consciousness, no longer instinctively.

If one wants to come in this way from the starry sky to a conception of space, then one does not get a space which encompasses three

---

1 The term "Imagination" here refers to a higher level of perception and consciousness in the sense of Rudolf Steiner's book: "Wie erlangt man Erkenntnisse der höheren Welten?" (How to gain knowledge of the higher worlds), GA 10.

dimensions, but one gets a space which I can only indicate pictorially as well: If I had to indicate the space of which I spoke yesterday by the three lines standing perpendicular to each other [in the middle of figure 1], I would have to indicate this other space in such a way that I would draw such configurations everywhere, as if forces in planes approached the earth from all sides of the universe and from the outside had a plastic effect on the formations which are on the surface of the earth.

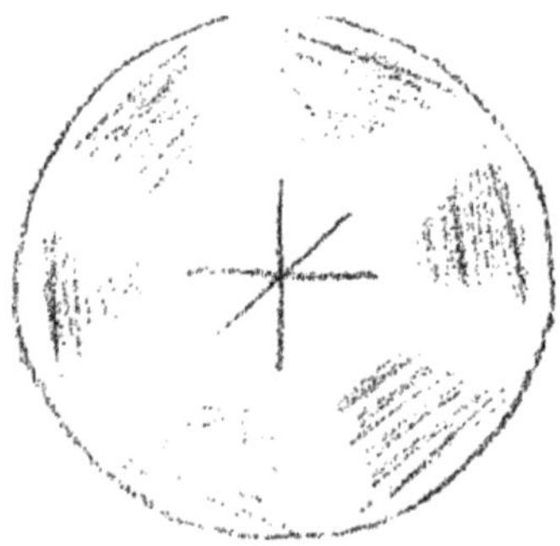

Figure 1:
Rudolf Steiner's sketch
of the formative forces acting in from the universe

One comes to such an idea when one advances from what can be seen with physical eyes at the living creatures, above all at the human being, to what I have now called here Imagination, whereby, instead of the physical human being, the cosmos reveals itself in image form and gives one a new space." [2]

---

[2] GA 82 "Damit der Mensch ganz Mensch werde" (So that the human being becomes completely human), The Hague, lecture of 9 April 1922.

This "new space", as Rudolf Steiner calls it here, was experienced by the ancients as a fourth dimension in addition to the three dimensions of the physical sense world. However, this pictorial experience by Imagination, was two-dimensional, because:

"It does continue with an indeterminate fourth, but one must turn around from a certain point, and the fourth dimension becomes namely simply the third dimension with a negative sign. [...] The fourth dimension is the negative third and annihilates the third, actually makes space two-dimensional." [3]

In addition to the cosmic, etheric formative forces, our emotional world also belongs to this "fourth dimension".

"In reality, the world of feelings is not spread out three-dimensionally, the world of feelings is in truth only spread out two-dimensionally." [4]

Therefore, it is impossible to perceive feelings with the conventional five senses which are directed to the three-dimensional external world. Nobody has ever seen a feeling with physical eyes, nor have they seen one of those forces of which the modern natural science speaks. It requires senses of the soul to be able to perceive soul contents or working forces. In the course of the last millennia these senses have been closed to us more and more. As a result, we have been placed completely outside the physical sensory world and today we tend to want to recognize this alone as reality. The mankind of older epochs would have denied such a view most violently, since this did not correspond to its life reality and world of perception.

---

3 Ibidem, lecture of 8 April 1922.

4 GA 213 "Menschenfragen und Weltenantworten" (Human questions and cosmic answers), Dornach, lecture of 24 June 1922.

From the twelve cardinal points of space completely different forces affect the earth and its inhabitants. With the help of the ancient clairvoyance one could distinguish these forces emotionally, experience and see them inwardly in a circle of twelve two-dimensional images. The physically visible star constellations in the sky were only later used as boundary markers between the images of the twelve force directions of space, felt and seen by Imagination. Rudolf Steiner explained about this in another lecture:

"You see, try now to take into account only, I would like to say, the spatial, by considering those spatial areas in the outer space which we can survey, which are just indicated to us in the outer world by what we call the zodiac. I do not want to pay a special attention to these zodiacal images in particular now. But take only the directions, to which we look, when we turn towards the constellation of Aries in the so-called zodiac, then to Taurus, Gemini, Cancer, Leo, Virgo, Libra, Scorpio, Sagittarius, Capricorn, Aquarius, Pisces. There we have to look, so to speak, first only at how the space, present to us as our visible universe, is divided. And only as a sign for this division we shall always point out to the region concerned; only as a sign in which direction we want to d e l i m i t the space, we shall refer to the relevant constellations in the zodiac.

Now the point is that these directions of space are really not something what one can characterize by saying: There is empty space and I draw some line into the empty space. Such a thing, which mathematics assumes as space, does not exist anywhere at all, but everywhere there are lines of forces, directions of forces, and these directions of forces are not equal, they are different from each other, they are differentiated. And one can distinguish these twelve areas of our visible universe by saying: If I look in the direction of Aries, the effect of force is different than if I look in the direction of Libra or in the direction of Cancer.

This is something, however, which at first man does not want to admit as long as he remains in the mere sense world. But at the moment when man ascends to the soul experience by Imagination, he does not feel indifferently the direction of Aries or Cancer, but he feels it in a highly differentiated way. [...] This graded, differentiated sensation is what the human being has when he ascends to Imaginative Perception, as soon as he moves around the firmament. These degrees of sensation, even these degrees of seeing, simply appear. This is the case at the moment when man comes out of the indifference of ordinary sense life. So we are not dealing here with something indifferent in space, but with the fact that the space around us affects us in a very differentiated way." [5]

Rudolf Steiner speaks of "Imagination" and "imaginative soul experience". He always used these terms to describe a level of perception and consciousness which makes impressions possible that lie in the realm of the fourth dimension or, as he usually calls it, in the "etheric region" or "etheric world".[6] In it, etheric formative forces work, which can be seen by "Imagination", i.e. in pictorial form, by people with sufficiently receptive souls. These forces affect the earth and its living beings from the periphery of the cosmos and give the vegetation process a completely different character in the course of the year. From month to month the appearance of the earthly

---

5 GA 201 "Entsprechungen zwischen Mikrokosmos und Makrokosmos – Der Mensch, eine Hyroglyphe des Weltenalls" (Correspondences between microcosm and macrocosm – Man: Hyroglyph of the Universe), Dornach, lecture of 17 April 1920.

6 E.g. in GA 26 "Anthroposophische Leitsätze" (Anthroposophical leading thoughts), chapter "Geistige Weltbereiche und menschliche Selbsterkenntnis" (Spiritual kingdoms and human self-knowledge), or GA 272 "Faust, der strebende Mensch – Band I" (Faust, the striving man – Vol. I), Dornach, lecture of 4 September 1916.

nature changes according to the zodiacal image or area of etheric forces the sun runs through at the respective time.

The twelve spatial directions, however, not only influence the annual vegetation process, but also the sensation in the souls of the people, changing in accordance with the course of the year.[7] Furthermore, the forces from the different directions of space even cause the successive ages and cultures of mankind by giving them each their very own character. Here, the so-called "vernal point" plays a special role. It is the point of intersection of the celestial equator with the ecliptic, which the sun passes annually at the beginning of spring (approx. 21 March). Opposite to it is the "autumnal point" (see figure 2). Both points are not fixed, but move, together with the slow rotation of the earth's axis, in about 72 years 1 degree backwards through the zodiac. This movement is called the precession of the vernal point.

Since the space and force area of a zodiacal image in the sky extends over a twelfth of the entire circle, thus covers 30° of 360°, the sun needs 30 x 72 years long, i.e. 2,160 years, until it has wandered backward through a single zodiacal image and has entered the influence area of the preceding one. For mankind this means the beginning of a new age and a new level of culture.

When did people begin to take an interest in such processes in the sky as the movement of the sun through the zodiac? According to Rudolf Steiner the common origin of astronomy and astrology was in a far past age, which began in the middle of the 6th millennium before Christ.

---

[7] See Rudolf Steiner's "Anthroposophical Soul Calendar" as well as the author's book "The Anthroposophical Soul Calendar and the incarnation cycle of man" (2020, Verlag BoD – Books on Demand, Norderstedt, Germany), in which it is described what can result from a longer meditative contemplation of the weekly verses of the Soul Calendar.

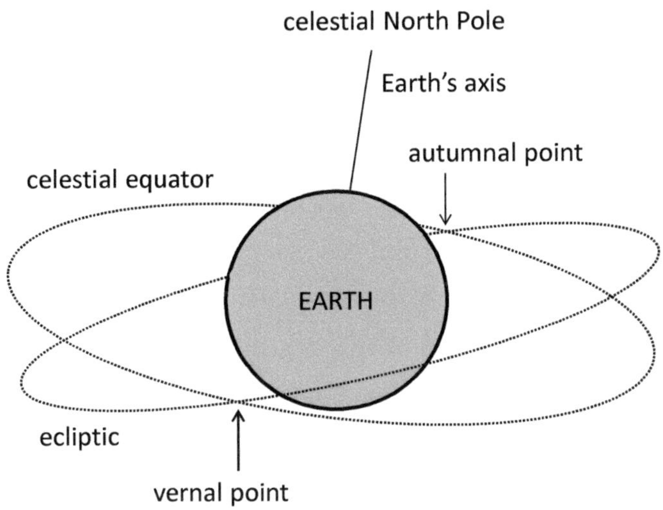

Figure 2:
The vernal point: intersection of the celestial equator with the ecliptic. [8]

Before this time the inner pictorial experience of humans was still so intensive, of such luminosity and liveliness, that the external sense perceptions appeared pale, shadowy and unreal in comparison. Therefore the inwardly experienced psycho-spiritual world was regarded as reality and the outer sense world only as an appearance, a Maya. Then, the vernal point moved through the zodiacal image of

8 According to Rudolf Steiner, the sun and the earth, as well as the other planets, move on lemniscatory paths. Only by the interaction of many different forces comes about what we describe today as the elliptical orbits of the planets. From the point of view of the lemniscatory paths the vernal point and the autumnal point, as intersections of the celestial equator with the ecliptic, are in the center of a double lemniscatory path of sun and earth. Detailed information about this can be found in the author's book "The Lemniscatory Path System" (2020, Verlag BoD – Books on Demand, Norderstedt, Germany).

Cancer. The people of that time lived entirely in their feeling. The most advanced ones founded a first post-glacial culture in India.

"This primeval Indian culture was far before the time in which the Vedas came into being. It still had something dreamlike, purely inwardly. The soul condition of the ancient Indian was completely opposite to ours today. To him everything external, visible was considered as Maya, as illusion, and the reality was only the Brahman[9] and what could be understood of the Brahman."[10]

It was not until the middle of the 6th millennium BC, when the vernal point moved backwards into the force area of Gemini, that to the west of India, in prehistoric Persia, the first appreciation of the outer sense world began. Physical outer world and psycho-spiritual inner world now stood opposite each other as a duality. Thus the experience of the people became dualistic. The opposites of light and darkness, day and night, past and future were topics that the cultural leaders of that time dealt with and taught their students about. Also earth and sky were now no longer experienced as a unity, but as a duality, both with physical senses and with the inner soul sense.

From the celestial direction in which then the sun rose at the beginning of spring, forces worked which caused images of the earthly human figure to arise in the soul, and this in a twofold way, as man and woman, as a human couple. This original representation of the zodiacal image, which was later called "Gemini" or "Twins", remained until the Middle Ages.

From the opposite direction of the sky, forces acted, which evoked the inner image of a psycho-spiritual man, residing in heaven, in the shape of a centaur with bow and arrow, which was later referred to

---

9 the soul or spirit of the world

10 GA 94 "Kosmogonie" (Cosmogony), Leipzig, lecture of 8 July 1906.

as "Sagittarius", "Archer" or "Shooter". His lower, animal part showed the influence of Ahriman, the god of earth and darkness, in the human soul. The upper, human part of the centaur was figurative expression of the influence of Ahura Mazdao or Ormuzd, the god of light and heaven. Under his guidance, man could rise above animalism.

The work of Ahura Mazdao found its external expression in all the light phenomena in the sky, in the course of the sun, the moon and the planets, as well as the shining stars in the night sky. Thus the greatest teacher of that age, the first Zarathustra, whose name his successors later carried on as a title, could found an initial, early astronomy:

"Now Zarathustra saw behind what is in the world of senses, especially in the way how the stars within the outer space join together to groups, such a sign writing. As we have the letters on the paper, so he saw in what surrounds us as the star world in the space something like the letters from the spiritual worlds which speak to us. And there was the art to penetrate into the spiritual world, and to read, to interpret the signs, which are given to us by the arrangement of the stars, to decipher from the movement and arrangement of the star worlds the way, how the spirits outside write their acts of the spirit-creation into the space. This became for Zarathustra and his disciples the thing which was important for them.

Thus, especially an important scriptural sign was this to them: that Ahura Mazdao accomplishes his creations, his revelations in the world by apparently having to describe a circle in the celestial space in the sense of our astronomy. This describing of a circle became the expression for a scriptural sign, which Ahura Mazdao or Ormuzd announces to the people, in order to show how he works, how he puts his deeds in the whole world context. There it was important that Zarathustra could point out: The zodiac is a line returning into itself, an expression for time returning into itself. In the highest sense of the word, one branch of time goes to the future, forward, the other to the past, backward. What later became the zodiac is Zaruana akarana: the time line finding itself in itself, described by Ormuzd, the Spirit of Light. This is the expression of the spiritual activity of Ormuzd.

The path of the sun through the zodiacal images is the expression of the activity of Ormuzd, and Zaruana akarana has its symbol in the zodiac. Basically, «Zaruana akarana» and «Zodiac» is the same word, like «Ormuzd» and «Ahura Mazdao». Two things are interwoven in the «going through the zodiac»: one is a certain course of the sun in lightness, when it sends its full forces as light forces on the earth in summer. But also its spiritual forces that it sends us from the light realm of Ormuzd. So, that part of the zodiac, which Ormuzd passes through during the day or during summer, shows us how Ormuzd works unhindered by Ahriman. Those zodiacal images, on the other hand, which lie below the horizon, symbolize the realm of shadow which, so to speak, Ahriman passes through." [11]

In addition, Zarathustra taught his students the structure of our planetary system, whereby each planetary sphere was regarded as the residence area of a certain group of spiritual beings. This geocentric view of the world, consisting of spheres of different sizes lying one inside the other, was later adopted by Ptolemy and further developed in a merely intellectual way. It even survived into the Middle Ages. It was not until the 16th century that it was gradually replaced by a heliocentric world view by Copernicus, as mankind was given the task of turning its attention for a time only to the physical sensory world. By limiting the perception to this, people were given the possibility to separate themselves from all external objects and

---

[11] GA 60 "Antworten der Geisteswissenschaft auf die großen Fragen des Daseins" (Answers of spiritual science to the great questions of existence), Berlin, lecture of 19 November 1911.

physically visible beings, and thus to develop a strong consciousness of themselves, a strong "I"-consciousness.

Fading of the Images

The Age of Gemini with the primeval Persian culture was followed next by the Age of Taurus. From the year 2907 B.C. the sun rose at the beginning of spring, for the duration of 2,160 years, in front of that celestial direction whose forces had evoked the image of a bull in the sensitive human souls of primeval times. Its counter-image in the sky was a large, eagle-like bird. In the course of the Age of Taurus it was transformed into a vulture, the bird of death, and in later times into the death-bringing scorpion. From its direction flowed to the people the power of spirit and life after death, from the opposite region of Taurus the power of substance and earthly life. Ankh, the Egyptian handle cross, became the most important symbol of the life force and the celestial bull was worshipped as the divine Apis. Murals vividly showed the sun giving life to the beings on earth by its rays.

Along with the increasing turn to the outer sense world with its manifold forms and bodies, however, the ability of the ancient clairvoyance faded away. One still knew about the pre- and post-existence of man in a psycho-spiritual world. But all perceptions in this respect faded noticeably, became more and more indistinct and shadowy. Thus the concern about life after death developed, which led to an ever more elaborately mortuary cult. Also, only a few persons, after special preparation in the mystery sites, were still able to experience those inner images when looking to the different directions of the sky, which were handed down to posterity as the twelve zodiacal images.

In order to be able to retain the knowledge of the boundaries between the twelve force areas, people looked for externally visible marks in the sky. For this purpose, the stars that are located close to the sun's path in the sky and physically visible to all people were grouped into twelve groups, the twelve constellations that we still call by the names of the originally clairvoyantly seen images. However, Rudolf Steiner points out that at that time these did not mark the space and force area of a zodiacal image, but that each constellation only indicates the direction where the boundary between two zodiacal images, seen by Imagination, is:

"In a sense, first we only have to look at how the space that is present to us as our visible universe is subdivided. And the region in question should only be pointed out as a sign of this subdivision; only as a sign in which direction we want to d e l i m i t space, are we referring to the constellations in question in the zodiac." [12]

The boundaries are therefore not between the physically visible constellations, as traditional astrology assumes, but somewhere in the inner area of them, since the constellations are large, broad directional markers. This knowledge was later lost. In the further development of the astronomical-astrological tradition, the view prevailed that the boundaries between the various zodiacal images or etheric force spaces, clairvoyantly seen by Imagination, were identical with the boundaries of the physically visible star constellations. However, there was the problem that these boundaries were not always clearly recognisable, since some of the physical constellations protrude into the area of the preceding or following ones, overlapping with them.

---

[12] GA 201 "Entsprechungen zwischen Mikrokosmos und Makrokosmos" (Correspondences between microcosm and macrocosm), Dornach, lecture of 17 April 1920.

Both factors together ultimately led to a shift in the spatial determination of the originally Imaginatively perceived twelve force areas. Already in the astronomy of the ancient Greeks, they no longer corresponded to the star groups and zodiac signs.

The Astronomy of the Ancient Greeks

In 747 B.C. the vernal equinox moved backwards into a force area that had evoked the image of a lamb or ram in the human souls of the preceding ages when they looked in the corresponding direction of the sky. The ability to see such inner celestial images had already been largely lost in the course of the Age of Taurus. However, the knowledge of these images was still very much alive. Even in the new age, people continued to cultivate this memory. It became part of the culture. In their consciousness humans distinguished themselves now much more from the world around them than it had ever been before.

The forces from the celestial direction of Cancer had directed people's attention entirely to their inner world of soul. Through the forces from the area of Gemini, people came to perceive the material outer world with the same intensity as their still very lively psycho-spiritual inner world. For the people of the Egyptian-Chaldean period, the physical sensory perceptions of the outer world already clearly outweighed the inner soul images, which were now fading.

Under the influence of the forces of the earth sign Taurus, one already learned to love the nature of the earth very much. All land was precisely measured and gigantic stone buildings were erected. Even in Babylonia, where the primeval Persian celestial science was continued as a tradition, huge stone observation towers were built, as it is handed down to us, for example, in the Bible as the Tower of Babel. The main focus here was on the world and the national community, less on the life of the individual.

Only with the entry of the retrograde vernal point into the force area of a new celestial direction did the attention of each every single person begin to be directed towards himself. Only now could a personality consciousness or "I"-consciousness develop. But the memory remained that the human being is transferred from the cosmos and a psycho-spiritual world to the earth and that the human form is built up from the feet to the head by the harmonious interaction of the forces from the twelve cardinal directions. The human body as the outer expression of the personality now became the focus of interest. Rudolf Steiner explained:

"If you see an ancient picture of a ram [see Figure 3], you will realise that it is not a naturalistic-materialistic picture of a ram, but the characteristic feature is always that the ram looks back, and that, the gesture, is the main thing. This looking back of the ram, that is the main thing about it. And this looking back of the ram, that is given in the looking back of the human being on himself, in this looking back on the universe that lives in him. So one must not just look at the ram in a naturalistic-materialistic way. It is not meant in this sense as an image, but the gesture of looking back is what matters." [13]

"When the sun entered the star image of the ram or lamb, the ram also appeared in the legends and myths of the peoples as something significant. Jason brings the ram's hide [the golden fleece] from Colchis. The Christ Jesus himself calls himself the Lamb of God, and he is represented symbolically, in the first period of Christianity, as the lamb at the foot of the cross." [14]

---

13 GA 208 "Anthroposophie als Kosmosophie – Teil 2" (Anthroposophy as Cosmosophy – Part 2), Dornach, lecture of 28 October 1921.

14 GA 54 "Die Welträtsel und die Anthroposophie" (Riddles of the World and Anthroposophy), Berlin, lecture of 12 April 1906.

Figure 3:
Rudolf Steiner's sketch
of the clairvoyantly seen image of a ram

Furthermore, Christ was portrayed as the shepherd who leads the lambs, i.e. the people of the Age of Aries. At the same time, he was experienced as the Saviour who brings everything back into balance that people have caused through wrongful individual actions. People should now also develop the ability to judge themselves by looking back at their own deeds and by examining to what extent their own actions serve or harm the common good. This weighing force flowed to them from that area of space which lies exactly opposite the image of a ram in the sky. It had evoked, by Imagination, the image of a pair of scales in the human souls of prehistoric times. Thus the conscience gradually emerged as an inner voice, for which a word has been there in the Greek language only since the Age of Aries.

"Here we can grasp with our hands, as it were, how the conscience is conquered for poetry. We see that Aeschylus [525 – 456 BC], the great poet, does not yet speak of conscience, while Euripides [c. 480 – 406 BC], his successor, already speaks of it. If we have this in mind, we can understand why human thinking, human earthly knowledge could only very slowly work its way up to a concept of conscience. The force that

works in the conscience was also at work in ancient times, when the images that represented the effects of human deeds were shown to clairvoyant vision." [15]

The images of avenging furies were now replaced by conscience. But the forces of Libra also enabled people to depict the human form in the highest beauty and harmony. How vivid and true-to-life the Greek human sculptures look compared to the rigid, caricature-like statues of the ancient Egyptians, Babylonians and Chaldeans. On the other hand, their knowledge of the cosmos and the forces of the twelve cardinal points, which was still vivid at the beginning and had been acquired through clairvoyant soul vision, had meanwhile solidified into mere tradition.

When Ptolemy wrote his astronomical Almagest and his astrological Tetrabiblos in the 2nd century AD, it had long since ceased to be possible to see the etheric zodiacal images as Imaginations. Seeing was replaced by an intellectual conception of the world. Thus it became the task of the Greco-Roman culture to penetrate the images and traditions, originally based on Imaginative visions, with the intellect and to assimilate them in a conceptual way. People were next to develop an intellectual soul.

Ptolemy based his works on the writings of Hipparchus of Nicaea (c. 190 – c. 120 BC), who died a few centuries before him. In his Almagest[16], he writes at the very beginning of the 7th book that *"Hipparchus' notes on the fixed stars, which we have preferably consulted for comparison, have come down to us in an impeccable version."* Hipparchus was apparently the first Greek ever to compile a

---

15 GA 59 "Metamorphosen des Seelenlebens – Pfade der Seelenerlebnisse" Teil 2 (Metamorphoses of soul life – Paths of soul experiences), Berlin, lecture of 5 May 1910.

16 Almagest, a major work of ancient Greek astronomy.

comprehensive star catalogue, for Ptolemy also informs us that Hipparchus *"found very few observations of the fixed stars made before his time, almost only those recorded by Aristyllus*[17] *and Timocharis*[18]*, and even these neither undoubtedly safe nor reliably edited."*

The Greeks apparently only began to practise scientific astronomy in the 2nd century BC. For this purpose, they drew connecting lines between individual stars in order to be able to clearly describe and find their respective positions. For example, Hipparchus gave the position of a star in the constellation called Triangle, located above Aries, in the following words: *"The preceding star (β Triangulum) on the base line of the Triangle deviates one inch east from a straight line which passes through the star (α) at the mouth of Aries and the left foot (γ Alamak) of Andromeda."*

This formulation shows that there was already a tradition at that time, according to which the images seen Imaginatively in the millennia before were converted into caricature-like line figures through purely mental and intellectual observation, now that these visions were no longer possible. Obviously, there were quite different views on this. Ptolemy, for example, was of the opinion that α Arietis, contrary to Hipparchus' opinion, was not at the mouth of Aries, but above its head. In his star catalogue, he therefore describes α Arietis with the words: *"The one above the head that Hipparchus places at the snout (or mouth)."*

The reproach of modern astronomy that the ancients projected highly imaginative images onto the sky is therefore at least partially justified, but only for the time when the direct vision of the zodiacal images by Imagination had long since ceased to be possible. At that

---

17 Aristyllus (around 261 BC), Greek astronomer

18 Timocharis of Alexandria (c. 320 - 260 BC), Greek astronomer and philosopher

time, the Greeks transferred numerous figures of their own mythology, such as Andromeda, Hercules, Perseus, Pegasus, etc., to groups of stars outside the zodiac and also gave individual stars the names of Greek heroes. By naming two bright stars close to each other after the twin brothers Castor and Pollux, they contributed to the fact that the originally Imaginatively seen zodiacal image of a human couple received the Greek name Δίδυμοι (Didimoi, Latin Gemini), i.e. twins. In this and similar ways, much of the ancient knowledge was changed and adapted to the imaginary world of the Greeks.

About the phenomenon of the shift of the vernal point there was still a great deal of uncertainty. Hipparchus first assumed that the constellations on the ecliptic were not real fixed stars, but mobile. Ptolemy reports that *"according to the first hypothesis put forward by Hipparchus, only the stars of the zodiac would move forward in the direction of the signs."* [19] So Hipparchus thought of the vernal point as fixed and instead of it the star groups of the zodiac as moving forward, whereby the vernal equinox only appeared to move backwards. At that time, no one knew that this movement was caused by a slow rotation of the Earth's axis, nor were they aware of the speed with which the vernal point shifted in relation to the zodiac. In his writing "On the Length of the Year", Hipparchus expressed the assumption based on a comparison of his own measurements with older measurements that had been handed down: *"If for this reason the solstices and the equinoxes receded at least 1/100° in one year against the direction of the signs, they must have receded at least 3° in 300 years."* Ptolemy, in the 2nd chapter of the 7th book of his Almagest, agreed with this assumption, according

---

[19] Almagest, book 7, chapter 1

to which the vernal point would only move back 1° in the zodiac after 100 years.

It was not until the 16th century AD that Copernicus recognised the movement of the Earth's axis as the cause of the precession of the vernal point.[20] However, on the basis of the data that had been handed down to him, he could only come to the conclusion that this movement was not uniform, but sometimes faster and sometimes slower. This is how inaccurate the measurement results of ancient astronomers were which have been preserved. Only more recent measurements in the 19th and 20th centuries were able to prove a movement of 1° per 72 years.

This clearly shows the great uncertainties with which ancient astronomy and astrology were afflicted. The lost possibility of the earlier clairvoyant, inner soul vision of the zodiacal images and, in addition, the ignorance of how far the vernal point had already moved backwards in relation to the constellations of the ecliptic in the previous centuries, brought enormous uncertainty into the astronomical-astrological world view of that time. In the end, one could only rely on fragmentary written heritage and a tradition that did not provide sufficient information about the observed shift in the starry sky. New knowledge had to be laboriously acquired.

In the star catalogue of the Almagest (7th and 8th book), Ptolemy assumed the position of the vernal point near the weakly shining star π Piscium (Pi Piscium).[21] Although it was counted among the constellation known as Pisces, Ptolemy gave 0° 10' ♈ as its ecliptic longitude. He saw that the star, according to its ecliptic longitude, was already in the region of the neighbouring constellation of Aries,

---

20 Nicolaus Copernicus, "De Revolutionibus", Book 1, Chapter 11, Nuremberg (Germany),1543.

21 "Piscium" is the genitive plural of the Latin word Pisces.

which was not unusual, since several of the physically visible constellations of the ecliptic overlapped. They were also of quite different sizes and therefore by no means congruent with the 30° wide traditional astrological zodiac signs of the same name.

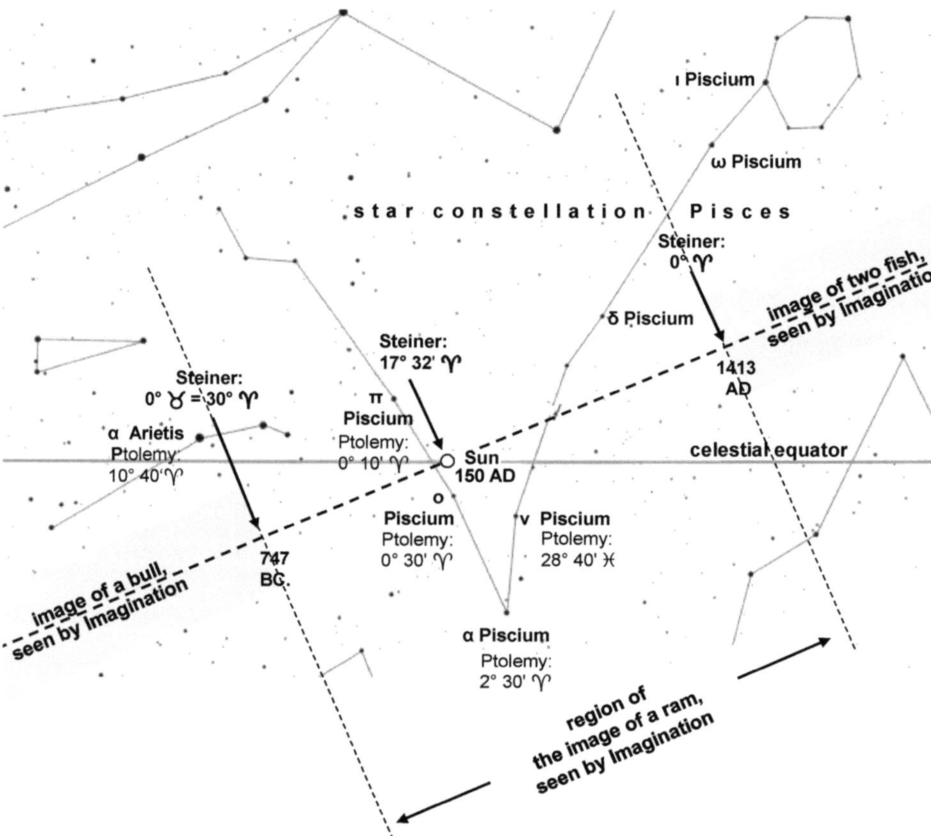

Figure 4:
Area of the Imaginatively seen image of a ram, which extended far into the area of the constellation later called Pisces (Pisces)

According to Rudolf Steiner, in 747 BC, at the beginning of spring, the sun moved into that region of the sky whose forces had once evoked the Imaginative image of a ram in the souls. At that time, the sun rose near α Arietis (Arabic Hamal = mutton or lamb), the main star of Aries.[22] Thus it is located near 0° ♉ or 30° ♈, so that this star could be used as a physically visible boundary marker in the sky between the force areas of Taurus and Aries. Ptolemy gave the position of α Arietis in his Almagest as 10° 40' ♈. With regard to this star, he thus deviated by 20° from Rudolf Steiner's information (see Figure 4).

Ptolemy lived from about 100 AD to 160 AD. He probably wrote the Almagest around the year 150 AD. Even if it had been a decade earlier, this would have no significance for determining the position of the vernal point during Ptolemy's lifetime, because it takes 72 years for it to move even 1 degree further. From the beginning of the Age of Aries in 747 BC to 150 AD, 747 + 150 = 897 years passed. If we divide this number by 72, we get the number of degrees that the vernal point moved backwards since the beginning of the Age of Aries until the year 150 AD in the area of the ram image, seen by Imagination. It is 12.4583°. Since the vernal point, coming from the area of the Imaginatively seen bull image, moved retrograde into the area of the ram image, we must subtract this number from 30° ♈ and thereby obtain 17.5417° ♈ or approx. 17° 32' ♈ as the position of the vernal point in the area of the once clairvoyantly seen ram image in 150 AD. In the star catalogue of the Almagest, Ptolemy gave 0° 10' ♈ and 0° 30' ♈ for the stars π Piscium and o Piscium, which were close to this position. In this case, his information deviates by

---

[22] Figures 4 and 5 were created with the astronomical freeware KStars. This was also used to calculate all the positions of the vernal point listed in this chapter.

more than half a zodiac sign from the position resulting from Rudolf Steiner's statements (see Figure 4).

If one determines the position of the vernal point for the year 2907 BC, with which, according to Rudolf Steiner, the age of Taurus preceding the age of Aries began, it can be seen that at that time the sun rose at the beginning of spring on the ecliptic above the brightly shining Aldebaran or α Tauri, the main star of the physical constellation of Taurus. This was obviously the visible boundary marker in the sky between the force area of the human couple or the twins, which the vernal point had just completely passed through, and the force area of the bull, into which the vernal point was now moving backwards.

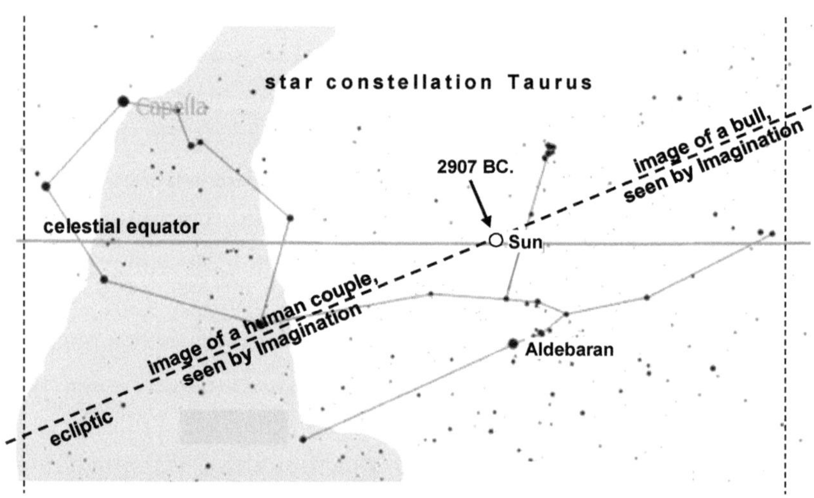

Figure 5: The sun in the vernal point above Aldebaran, the main star of the constellation known as Taurus

(The grey shadow on the left represents the Milky Way.)

If one calculates the positions of the vernal point in steps of 2,160 years each for the entire zodiac, there is no noticeable proximity to a particularly bright star of the Gemini and Cancer constellations for the beginning of the ages of Gemini and Cancer. However, the brightest stars of Gemini, which the Greeks named Castor and Pollux, are located at the very edge of this star group, and the constellation of Cancer contains no conspicuously bright stars.

The neighbouring constellation of Leo, on the other hand, presents itself in the night sky with the brightly shining Regulus or α Leonis. Together with it, in close conjunction, the Sun rose at the beginning of spring in the year 9387 BC. Regulus marked the boundary to the preceding Imaginative zodiacal image of Virgo.

Similar conspicuous stars are found for the borders Sagittarius/Scorpio with the brightly shining Antares or α Scorpii, for Capricorn/Sagittarius with ε Sagittarii, the brightest star of the group, although it is not marked with the letter α, for Aquarius/ Capricorn with the stars α^2 and β Capricorni, which are close to each other and still the second brightest, and for Pisces/Aquarius with α Aquarii, also the second brightest star. Like α Arietis (Hamal) and α Tauri (Aldebaran), they apparently served as boundary marks in the sky, visible to physical eyes, with which people in the preceding Egyptian-Babylonian age, when ancient clairvoyance was increasingly fading, had marked the transition from one force area of the zodiac to the next.

In the constellation Virgo, the boundary between the Imaginative celestial images and force areas of Libra and Virgo falls more in the initial area (near η Virginis), while the brightly shining Spica (α Virginis) is more in the final area of the star group. Since this group is very elongated, the boundary of the celestial image of Libra to Scorpio is still in the area of the constellation Virgo. It is close to ι (iota) Virginis and thus closer to Spica than to the first star of the Libra group.

Of the Imaginatively seen zodiacal image of Pisces, the boundary marker to the preceding force area of Aries was already indicated as lying between the inconspicuous stars δ and ω Piscium (see Figure 4, p. 30). In 2020, the vernal point had already moved on to near the star ι (iota) Piscium.

In summary, for the force areas Aries, Taurus, Leo, Scorpio, Sagittarius, Capricorn and Aquarius, brightly shining stars can be found in the sky as boundary marks to the respective preceding sign. This is not the case with the force areas of Gemini, Cancer, Virgo, Libra and Pisces. Altogether there are seven bright boundary marks and five dark ones. This ratio corresponds to an occult law according to which the ratio between light and darkness in the cosmos is generally 7:5. In the zodiac this is also expressed in other ways, for example in the division into seven light and five dark zodiacal images, as Rudolf Steiner explained:

"so you have: Aries, Taurus, Gemini, Cancer, Leo, Virgo, Libra; seven celestial signs for the light side. Five for the dark side: Scorpio, Sagittarius, Capricorn, Aquarius, Pisces. Day, night." [23]

[23] GA 183 "Die Wissenschaft vom Werden des Menschen" (The science of the becoming of man), Dornach, lecture of 25 August 1918.

Consequences for Astrology

The first step in creating a horoscope is to calculate and draw it. When Ptolemy wrote his astrological work Tetrabiblos, he was in agreement with Hipparchus and the astronomical-astrological tradition of his time that the twelve Imaginatively experienced zodiacal images, handed down through the centuries, would coincide with the physically visible star constellations. The rising of the sun at the beginning of spring in the border area of the constellation Pisces fitted perfectly with the idea that there was 0° ♈. Accordingly, the vernal point had become the "point of Aries" already in the early days of Greek culture.

Later Western astrology retained this conception of the zodiac as the so-called "tropical zodiac" and fixed the vernal point unchangeably at 0° Aries, followed by the series of the twelve zodiac signs. The term "tropical" refers to the equinoxes and solstices. Eastern astrology, on the other hand, takes the physically visible constellations, the so-called "sidereal zodiac" as the basis for calculation. At the time of Ptolemy, both were still essentially in agreement, because for the astrologers of that time there was initially no reason to take the precession movement of the vernal equinox into account. Its position finally changed by just 1 degree only after 72 years. If they had wanted to take this shift into account, they would have miscalculated anyway, since the successors of Hipparchus and Ptolemy would certainly have followed their assumption that it took 100 years for the vernal point to move through 1 degree.

Another unknown fact was that the twelve celestial sections of the tropical zodiac as well as those of the sidereal zodiac did not correspond to the positions of the Imaginatively seen celestial images, which were an expression of the very differently working forces of the twelve spatial directions. The Greeks themselves could no longer see any Imaginative celestial images. They had to rely on the little information from the Chaldean-Babylonian tradition that was still available to them. However, even then these contained only fragments and also already falsifications of the original knowledge. Similarly, the original knowledge about the reincarnation of humans, which had survived into the time of the primeval Persian culture, had been distorted in the course of the Egyptian-Chaldean-Babylonian culture into the doctrine of the "transmigration of souls", according to which humans would also incarnate in animals, even of the lowest level. Thus it was unavoidable that much that was false and untruthful was passed on from the third to the fourth culture, both with regard to astrology and with regard to the doctrine of rebirth. Rudolf Steiner's judgement in this respect is:

"Such an echo of ancient wisdom, which today's natural scientists look down upon with a shrug of the shoulders, has also been preserved in what is called astrology in a mutilated, foolish way, but nevertheless leads back to the primal wisdom of mankind." [24]

Accordingly, Greek astrology had early on become an only partially correct and, unfortunately, moreover, quite immobile, rigid object. It fell victim to the general tendency of all tradition-forming to cement some realities, negate others and disregard necessary changes. Nevertheless, the twelve force areas coincided at least partially with the signs of the zodiac assumed by Hipparchus and

---

[24] GA 56 "Die Erkenntnis der Seele und des Geistes" (Knowledge of soul and spirit), Berlin, lecture of 26 March 1908.

Ptolemy in those days, so that we may assume that there were always cases in which astrological predictions proved true. What Rudolf Steiner said about Kepler's time at the end of the Middle Ages was probably already valid for the ancient Greeks and is certainly just as valid today:

"Certainly, in an infinite number of cases, it must be admitted that the striking of the impression one may have of astrological predictions that come true is simply due to the fact that one is surprised by the occurrence of such a prediction and retains what agrees and forgets about what did not come to pass."[25]

Anyone who, like the author of this book, has seriously studied astrology for decades will have to admit, on thorough soul-examination, that the tendency to brush aside predictions that did not come true, on the other hand to joyfully keep in one's memory predictions that did come true, is very great. In the Middle Ages, moreover, in order to make things fit, additional birth time corrections were carried out, which, however, did not at all please the famous Kepler, who, like many astronomers of his time, was repeatedly asked for astrological information. Rudolf Steiner commented on this event with the words:

"It must be said, however, that Kepler was a thoroughly honest man and did not like to do anything like correcting the hour of birth. From the letter that Kepler wrote to Wallenstein about this at the time, one senses that he did not like to do it and could not recommend such a procedure, because if one undertakes something like this, one could establish everything possible in this way."[26]

---

25 GA 61 "Menschengeschichte im Lichte der Geistesforschung" (Human history in the light of spiritual research), Berlin, lecture of 9 September 1911.

26 Ibidem.

According to Rudolf Steiner's statements, the vernal point had wandered backwards through the entire range of the Imaginatively seen ram image by the year 1413 and had arrived at 0° ♈. Only now had it really become the "Aries point". In this way, in the 15th century, the astrologically effective force areas of the originally Imaginatively seen zodiacal images came to coincide with the zodiac signs of the tropical zodiac. This was of course to the benefit of the medieval astrologers, even though they knew nothing of this process. Until the middle of the 16th century, the vernal point moved about 2° backwards into the force area of Pisces, i.e. up to 28° Pisces. This still resulted in a much more favourable situation than existed for the ancient Greeks. Their tradition was continued unwaveringly and, when listing the daily planetary positions in the ephemerides, the "tropical" zodiac with its immovable vernal point was retained. The position of the Sun at the beginning of spring on 21 March was always given as 0° ♈, although it was still rising in the power area of Pisces and only reached that of Aries two days later.

With increasing temporal distance to the year 1413, the temporarily existing correspondence of the force areas, which had been seen clairvoyantly and were effective in the sense of astrology, with the signs of the tropical zodiac decreased again. When drawing up a horoscope, the precession of the vernal point should therefore always have been taken into account, for it soon shifted considerably further retrograde into the force area of Pisces. Figure 6 shows the resulting shifts for the time after 1413 in tabular form.

Regarding the birth of a human being in 1917 at the beginning of spring on 21 March, one finds in the ephemerides, which are based on the immovably rigid model of the tropical zodiac of traditional Western astrology, the position of the Sun and thus of the vernal point 0° Aries. In fact, however, both were still in the astrologically effective force space of Pisces, at 30° minus 7°, i.e. at 23° ♓. A person born on 21 March 1917 was therefore astrologically not "Aries" at all, but "Pisces". For people born in 1953 the deviation was already 7.5°. The vernal point had moved back to 22.5° ♓. For those born in 1989 8°, and for those born in the 2020s even 8.5° of shift have to be taken into account, because the vernal point has meanwhile reached about 21.5° ♓.

1413	0°	0° ♈	1809	5,5°	24,5° ♓
1449	0,5°	29,5° ♓	1845	6°	24° ♓
1485	1°	29° ♓	1881	6,5°	23,5° ♓
1521	1,5°	28,5° ♓	1917	7°	23° ♓
1557	2°	28° ♓	1953	7,5°	22,5° ♓
1593	2,5°	27,5° ♓	1989	8°	22° ♓
1629	3°	27° ♓	2025	8,5°	21,5° ♓
1665	3,5°	26,5° ♓	2061	9°	21° ♓
1701	4°	26° ♓	2097	9,5°	20,5° ♓
1737	4,5°	25,5° ♓	2133	10°	20° ♓
1773	5°	25° ♓	2169	10,5°	19,5° ♓
			2205	11°	19° ♓

Figure 6:
Shifts and positions of the vernal point since the year 1413

This shift naturally affects not only the positions of the sun and the vernal point in the zodiac, but also those of all the other planets. Accordingly, the number of degrees listed in the above table would always have to be subtracted from the positions given in the ephemerides for the respective year of birth. Sun and planets would then have to be entered in the horoscope shifted backwards by this distance. If the drawing is created and printed by a computer programme, the border lines between the zodiac signs can alternatively be entered with a pencil, shifted by the corresponding number of degrees. For those born around the year 1990, the border line to the previous sign would be entered in 8° of each sign. Thus, for example, the power range of Pisces would extend to 8° Aries. 0° Aries in the prefabricated horoscope drawing would then correspond to 23° Pisces. The power range of Aries would accordingly extend to 8° Taurus, the power range of Taurus to 8° Gemini, etc.

Since each degree in the zodiac corresponds to about one day of the sun's course, the astrologically effective force area of Aries no longer begins on 21 March, as it is handled according to the traditional "tropical" rule, but only 8 days later, on 29 March, due to the shift by 8° according to the above table. The result is that today the astrologically effective force spaces of the zodiacal images almost coincide with the calendar months. Aries now begins just two days before April 1. Its force area has already become almost congruent with the month of April. It will even become more and more so in the course of the next two centuries. In 2205, the vernal point will have shifted by 11° from 0° Aries, so that Aries will then only begin 11 days after 21 March: on 1 April.

The previous failure to take this into account in traditional astrology may be the reason why one can occasionally meet persons who assure you that the astrological statements about their alleged zodiac sign definitely do not apply to them, but that they find themselves quite well described by the characteristics of the preceding sign.

However, creating a horoscope drawing is only the first step of an astrological consideration. It is merely the basis for the second step: the reading of the horoscope. Here the loss of original knowledge is probably even greater than in the determination of the spatial areas of the zodiac forces, since many rules of reading, such as those described by Ptolemy in his Tetrabiblos, were only a late product of the formation of tradition, frozen and distorted into a caricature. Rudolf Steiner's severe judgement, according to which the original knowledge *"has been preserved in what is called astrology in a mutilated, foolish way"* [27], therefore seems fully justified, not only because the capacity for inner soul vision by Imagination had long been lost, but also because, in order to truly interpret a horoscope, the human soul must develop the last and highest level of the three higher levels of consciousness of Imagination, Inspiration and Intuition[28], which is the most difficult to attain, and which is nowadays possible only for very few people in the context of a longer initiation process.

"Real astrology, however, is a completely i n t u i t i v e science and requires the development of higher psychical powers of cognition in those who wish to practise it, which can be present in v e r y f e w people today. [...] There have been and still are persons in the secret schools who can practise astrology in this sense. And what is written about it in the accessible books has in some way come from such secret teachers. But everything that deals with these things is inaccessible to the common mind, even if it is written in books. For in order to understand them, a deep I n t u i t i o n is needed. And what has been copied from the real compilations of the teachers by those who have not understood it themselves, is of course not really suitable to give a favourable opinion of astrology to people who are caught up in the present way of thinking. But it must be said that even such books on astrology are not entirely worthless. For people copy all the better the

---

27 GA 56 "Die Erkenntnis der Seele und des Geistes" (Knowledge of soul and spirit), Berlin, lecture of 26 March 1908.

28 The terms Imagination, Inspiration and Intuition are used here as expressions for three higher levels of consciousness in the sense of Rudolf Steiner's explanations in his book "Wie erlangt man Erkenntnisse der höheren Welten?" (How to gain knowledge of the higher worlds?), GA 10.

less they understand what they copy, they do not then spoil it by their own wisdom. Thus it comes about that in astrological writings, however dark their origin, pearls of truth are always to be found for him who is capable of Intuition – but only for such a one."[29]

Which astrologer of our time can claim to fulfil these requirements? Astrological readings today can only be cautious attempts of interpretation and even in the best of cases will only be able to produce fragments of the truth. On the other hand, the temptation will be great to use rules that *"have been preserved in a mutilated, foolish way"* and even to create additional rules if necessary so that the result may better satisfy expectations. To this end, a whole series of different astrological "schools" have emerged in recent centuries, each of which is of course convinced that it can provide the best and most accurate results in creating and reading of horoscopes.

[29] GA 34, Journal "Lucifer-Gnosis", 1903 – 1908, p. 396, on the question "How does theosophy [former name for anthroposophy] relate to astrology?"

Duration of the Ages

In his many lectures, Rudolf Steiner occasionally gave information on the duration of the ages in connection with the rotation of the vernal equinox, i.e. the point of sunrise at the beginning of spring, through the entire zodiac. One of his statements in this regard reads, for example:

"The sun makes a revolution. In 25,920 years it makes a revolution around the whole world." [30]

This revolution is called a "Platonic Year" or a "Great Year". Assuming a steady movement of the vernal point through the entire zodiac, this results in one twelfth of the stated time for a single sign of the zodiac. That is 2,160 years. In the same lecture Rudolf Steiner gave the following explanation:

"So when we go back 2,160 years, it rose in Aries; when we go back another 2,160 years, it rose in Taurus, another 2,160 years back in Gemini; and another 2,160 years back in Cancer. [...] But if we go into the future now, we shall have the Sun in Libra once again; for now it rises in Pisces, after 2,160 years in Aquarius, then in Capricorn, in Sagittarius, in Scorpio, and once again in Libra."

On another occasion he said:

"Then comes a fourth age that developed in the south of Europe, the age of the Greco-Latin culture. [...] The Roman world is also part of this.

---

30 GA 349 "Vom Leben des Menschen und der Erde – Über das Wesen des Christentums" (On the Life of Man and Earth – On the Essence of Christianity), Dornach, lecture of 17 February 1923.

It is an epoch that begins around the 8th century, 747 BC, and lasted until the 14th and 15th centuries, 1413 AD. From then on we have the fifth period in which we ourselves live." [31]

From 747 BC to 1413 AD it is exactly 2,160 years. This and the preceding quotation could lead one to believe that an age or culture always lasts 2,160 years. In a lecture from 1920, Rudolf Steiner described the beginning and end of the five post-Atlantean (post-glacial) cultures more precisely with the following dates: [32]

I	Primeval Indian	from 8167 until 5567 BC
II	Primeval Persian	from 5567 until 2907 BC
III	Egyptian-Chaldean	from 2907 until 747 BC
IV	Greco-Latin	from 747 BC until 1413 AD
V	Now	from 1413 until ...

Only for the Egyptian-Chaldean culture and the Greco-Latin culture, which correspond to the Taurus and Aries ages, was a duration of 2,160 years given here. In the case of the previous primeval Persian culture, the Age of Gemini, it is 5567 - 2907 = 2,660 years. For the even older primeval Indian culture it is 8167 - 5567 = 2,600 years.[33] Rudolf Steiner did not comment on the exact duration of our fifth culture or the Age of Pisces. The numbers he himself wrote on the blackboard could at first glance be perceived as

---

31 GA 106 "Ägyptische Mythen und Mysterien" (Egyptian myths and mysteries), Leipzig, lecture of 2 September 1908.

32 GA 196 "Geistige und soziale Wandlungen in der Menschheitsentwicklung" (Spiritual and social transformations in human development), Dornach, lecture of 16 January 1920.

33 In early lectures Rudolf Steiner gave 2,600 years as the duration of an age, e.g. in GA 93a "Grundelemente der Esoterik" (Basic elements of Esotericism), Berlin, lectures of 3 and 8 October 1905. Later he always gave 2,160 years as the average duration of the ages.

a contradiction to his statements quoted above, where he always speaks of 2,160 years.

But, the contradiction quickly clears up, if one takes into account that Rudolf Steiner was calculating there from the year 1923 when he said: *"when we go back 2,160 years [i.e. from the year of the lecture], it rose [the sun at the beginning of spring] in Aries; when we go back another 2,160 years, it rose in Taurus, ..."* and *"for now it rises in Pisces, after 2,160 years in Aquarius, then in Capricorn, ..."*

If we subtract 2,160 years from the year 1923 AD, we get the year 237 BC. It lies within the period of 747 BC to 1413 AD given for the fourth culture, the Age of Aries. If one subtracts 2,160 years from the year 1923 several times, one always obtains dates that lie between the beginning and the end of the cultures mentioned:

1923 AD – 2,160 years = 237 BC	Aries	culture IV: 747 BC until 1413 AD
237 BC – 2,160 years = 2397 BC	Taurus	culture III: 2907 BC until 747 BC
2397 BC – 2,160 years = 4557 BC	Gemini	culture II: 5567 BC until 2907 BC
4557 BC – 2,160 years = 6717 BC	Cancer	culture I: 8167 BC until 5567 BC

Thus there is no contradiction between Rudolf Steiner's calculation of 2,160 years each and the concrete years he specified for the duration of the individual cultures or ages.

A completely different problem arises from this, however. If the vernal point requires 12 x 2,160 years, i.e. a total of 25,920 years for a complete revolution through the zodiac, but two ages lasted longer than 2,160 years, then consequently other ages must be shorter by exactly the same amount of time that the other two are longer. The Gemini Age lasted the longest at 2,660 years. The preceding Cancer Age was slightly shorter at 2,600 years.

This raises the question of the cause of the different duration of the ages. Rudolf Steiner explicitly connects the zodiacal forces with the twelve spatial directions, as explained in detail in Chapter 1. His sketch shown as Figure 1 (p. 10) accordingly contains six shaded areas of equal size, opposite each other, representing the twelve force areas of the space surrounding the earth. According to this, one should be able to assume that these areas of force are all of the same size. Independently of Rudolf Steiner's sketch, we also understand spatial directions as an even subdivision of a circle encompassing 360°. This is probably why Rudolf Steiner usually calculated with the average number of 2,160 years. The assumption that those two spatial directions, whose forces evoked the Imaginative images of a cancer and a human couple (twins) in the human souls of earlier times, would occupy a wider space in the sky than the others, therefore seems rather unlikely.

Nor can the different extents of the constellations Gemini and Cancer be held responsible for the different duration of the Gemini and Cancer ages, since the actual force areas of the zodiac are not at all identical with the physically visible constellations, as has been explained in detail in the previous chapters. Furthermore, the constellation of Cancer in the sky belongs to the smaller star groups rather than to the larger ones. This could at best justify a smaller, but in no case such a large duration of the age of Cancer as Rudolf Steiner indicated. The same applies to the constellation Aries. It too has only a small extension in the sky. Much larger is the neighbouring constellation of Taurus. Nevertheless, Rudolf Steiner gave exactly the same duration of 2,160 years for the Age of Taurus and the Age of Aries.

In principle, it would be conceivable that each of the twelve areas of twelve power areas of the zodiac that are invisible to the physical eyes of its own magnitude, as is the case with the physically visible star constellations. But then each power area would have to have its own individual size. This is obviously not the case, because according to Rudolf Steiner's statements, the power areas of Taurus and Aries are exactly the same size, since the vernal equinox took 2,160 years to pass through them backwards in both cases.

In principle, it would be conceivable that each of the twelve force areas of the zodiac, which are invisible to physical eyes, has its very own size, just as is the case with the physically visible star constellations. But then each power area would have to have its very own individual size. This is obviously not the case, because according to Rudolf Steiner's statements, the force areas of Taurus and Aries are exactly the same size, since the vernal point took 2,160 years to pass through them backwards in both cases.

Independently of this, a greater extension of the force areas of Gemini and Cancer would shift the beginning of Leo and the subsequent zodiacal images by a good half of a force area. For the Gemini Age, which lasts 2,660 years, a power range of 37° would have to be assumed instead of 30°, and for the Cancer Age, which lasts 2,600 years, a power range of 36° would have to be assumed.[34] 7° plus 6° equals 13°. In addition, due to the precession-related shift of all twelve force areas, there would be an additional 8°, as described in detail in the previous chapter. Ultimately, this results in a total shift of 21°. Astrologically, this would result in two-thirds of the traditional zodiac sign Leo still belonging to the force area of Cancer, likewise two-thirds of the zodiac sign Virgo still belonging to the zodiac sign Leo, and so on until one or more force areas would follow which would be considerably narrower than the average of 30° to compensate for the greater expansion of Gemini and Cancer. In

---

[34] Gemini Age: 2660/2,160 = 1.23 und 1.23 x 30° = 36.94°.
Cancer Age: 2600/2,160 = 1.20 und 1.20 x 30° = 36.09°.

the end, there must not be more than 12 x 2,160 = 25,920 years, since, according to Rudolf Steiner, the vernal point travels through the entire zodiac during this time. Such a strong shift seems rather unlikely and cannot be confirmed by horoscopes.

But if the spatial areas of the twelve zodiacal forces are not different in size, then a different duration of the ages is only possible if the vernal point moves through the zodiac at different speeds. Rudolf Steiner's statements on the duration of the ages would thus have consequences for astronomy. This will be examined in more detail in the following.

Consequences for Astronomy

Vernal point and autumnal point are the two intersections of the celestial equator with the ecliptic. This is clearly shown in Figure 2 (p. 15). The earth's axis is perpendicular to the centre of the plane of the celestial equator. If it changes its position, it has a direct effect on the celestial equator. If the earth's axis rotates in a circle, the two points of intersection move with it. Vernal point and autumn point then move along the ecliptic accordingly. How fast both points of intersection move depends on the speed at which the earth's axis rotates.

When Copernicus first recognised this connection in the 16th century and explained it in his book "De Revolutionibus"[35], he based on the historical data available to him, in particular those provided by Ptolemy in his Almagest. For this reason, he could only come to the conclusion that the rotation of the Earth's axis, which causes the backward movement of the vernal point through the zodiac, occurs at different speeds. Copernicus assumed that his predecessors had determined their measurements with all due care and that he could therefore trust them. Only modern astronomy was able to prove that this trust was not justified. One of the reasons for this was certainly the difficulties that astronomical measurements are fundamentally associated with. The accuracy of the measurement results depends not only on the care of the observer, but also very much on the quality of the instruments used. The ancient and medieval

---

35 Nicolaus Copernicus, "De Revolutionibus", book 1, chapter 11, Nuremberg, 1543.

astronomers either had to make them themselves or have them made by a craftsman according to their specifications. This is a major source of inaccurate measurement results.

Today, astronomy assumes that the vernal equinox moves 1° backwards in the zodiac in 72 years. Rudolf Steiner also holds this view, at least as far as the last five thousand years of human history are concerned. For the Taurus Age or the Egyptian-Chaldean culture and for the Aries Age or the Greco-Latin culture, he gives a duration of 2,160 years each. If we assume that the force areas of Taurus and Aries each extend over 30° of the entire zodiac, we get 2,160 years divided by 30 = 72 years for the time it takes the vernal point to travel through 1° of the zodiac.

Today, no astronomer can say with sufficient certainty whether the speed with which the Earth's axis rotates is really constant. We do not have any data from pre-Egyptian or pre-Babylonian times, and even if this were the case, their reliability would have to be very much questioned. All statements of modern astronomy about the positions of the vernal point in earlier cultures of mankind are therefore based solely on back-calculations using the current speed of rotation of the Earth's axis, which is considered to be constant.

In reality, however, today's knowledge about the movements observable on Earth as well as on the other planets is still very incomplete. Such a statement may come as a surprise, since we are in the age of space travel and send probes to planets and asteroids that undoubtedly reach their destination. However, this is based on a completely different set of facts. In the case of spacecraft and space probes, scientists have a very precise overview of the forces at work in them, because they have to develop and construct everything themselves. But this is by no means the case with the planets. We find them "completely constructed" and can only try to find out and understand through observation how and why they behave the way they do.

Ever since Kant and Laplace[36] published their theories on the creation of our planetary system, the belief has prevailed that its formation is based solely on a centripetal force of attraction, gravity, and a centrifugal force. The problem here, however, is that neither a force that pulls everything towards a centre nor a force that acts in the opposite direction can bring about a movement that is at right angles to both directions of force. Bodies that are only subject to these two forces can only either move towards the centre or away from it, depending on which of the two forces predominates. However, it is impossible for them to move around a centre point in circular, elliptical or any other kind of path.[37] This requires a force that acts in a completely different direction.

Also, a body that is exposed to the two centripetal and centrifugal forces would have no reason whatsoever to set itself in rotation around its axis. On the contrary, science explains the very slow rotation of the moon (c. 27.3 days), which always turns the same side towards the earth, precisely with the earth's gravitational effect. According to this, gravity slows down every rotation and in no way causes it. Neither the course of the planets around the sun nor their rotation can therefore be explained with the help of Kant-Laplace theory. Obviously there are other forces at work here that are still completely unknown to today's science.

Everything we know about the movement of the Earth's axis is based on observations and measurements of the last few centuries.

---

36 Immanuel Kant (1724–1804), German philosopher, and Pierre-Simon Laplace (1749–1827), French mathematician, physicist and astronomer.

37 Rudolf Steiner ascribes a lemniscatory movement to the planets. (see footnote 8, page 15).

We do not have any reliable data about the behaviour of the Earth's axis a few millennia ago. It is therefore quite possible that its rotational speed is not constant over long periods of time, but is sometimes faster and sometimes slower. If this were the case, the longer duration of the Gemini and Cancer Ages as stated by Rudolf Steiner could be explained by a slower movement of the Earth's axis with a resulting slower precession of the vernal point.

The result described at the end of the chapter "The Astronomy of the Ancient Greeks", according to which some particularly bright stars of the night sky have served since antiquity as physically visible boundary marks between some forces spaces of the zodiac, remains unaffected by the possibility of a slower movement of the Earth's axis. A distance of 30° in the zodiac would then be covered in more years. In the end, however, there would still be 30° to be covered. Those bright stars of the night sky, which represent boundary marks, could thus continue to fulfil their tasks. Of course, however, we would have to assume a gradually increasing deceleration of the rotation of the Earth's axis, presumably beginning sometime in the Cancer Age into the Gemini Age, and a subsequent gradual acceleration, so that by the Taurus Age the average 2,160 years could be reached again. This would have to be balanced by a reverse process in the passage of the vernal point through the force spaces of Capricorn and Sagittarius, which are opposite Cancer and Gemini in the zodiac: by a gradual acceleration followed by a gradual deceleration.

At first glance, such a change in speed may seem unlikely. However, if one takes into account that not only in terrestrial but also in extraterrestrial nature many processes run rhythmically, such as the faster movement of the planets near the sun and their slower movement farther away from the sun, then the idea of a rhythmic change in the speed of the rotation of the earth's axis may appear less absurd.

We do not even know today whether the rotation of the elliptical orbits, the so-called perihelial or apsidal rotation, which is most conspicuous in the Mercury orbit, but to a lesser extent in all planetary orbits, is constant in speed or whether it also follows a rhythm. Likewise, the distances of the planets from the sun could be subject to rhythmic fluctuations over a period of millennia. There is still much that is unclear about planetary motion, including, as already mentioned, the very fundamental question of which force actually drives the planets in their orbits and which other force causes them to rotate around their axis. The question of why the orbits of the planets have different inclinations will also have to wait a long time for an answer. Obviously, many more forces are involved in the structure and processes of a planetary system than modern natural science has been willing to admit so far.

Influences of the Forces of the Zodiac on the Cultural Development of Europe

The possibility of an influence of the zodiac forces on the cultural development of mankind can only be denied by modern natural science and dismissed as superstition, since it already denies the existence of such forces in general. Space is imagined to be equal in all directions and not qualitatively differentiated with regard to the forces at work in it. From a scientific point of view, it is not yet comprehensible that processes in the cosmos, such as the precession of the vernal point in the zodiac, could have not only a physical but also a psycho-spiritual influence on humanity, simply for the reason that one only recognises the Physical as objective reality anyway and classifies everything Psycho-Spiritual as subjective and therefore unreal.

Nevertheless, no one can deny that the peoples of antiquity and preceding times show clear characteristics in their cultures that point to such an influence. The veneration of the bull in the third culture alone, which not only found expression in the Apis cult of the Egyptians or in the Mithras cult of Asia Minor, but even today shows its after-effects in the veneration of cattle as sacred animals in India or the tradition of bullfighting in Spain, are unmistakable indications of an influence of the passage of the vernal equinox through the spatial area of that zodiac force which evoked the image of a bull in the souls of the people of prehistoric times. The opposite zodiac force, which at the same time acted as an antipole, was originally seen figuratively as an eagle-like bird. The massive bull and the soaring eagle were and are imaginative expressions of the opposition between earthly substance and heavenly spirit. Man oscillates back and forth between these two principles in his incarnational cycle. If he is completely spirit, he is bodiless, i.e. without substance. As the eagle rises from the earth to the sky, so the spirit rises from the substance at death. Then the soul and spirit of the deceased ascend to heavenly worlds.

Due to this connection between spirit and death, the image of the eagle in ancient Egypt changed to that of a vulture, which feeds on dead material and circles up in the sky in the light of the sun. In later traditions, the vulture turned into the deadly scorpion. This is why the cult of the dead played such an important role in ancient Egypt. We can only understand this culture at all if we look at it in the dichotomy of its turning to the material outside world, the mastery of earthly forces (bull) on the one hand and its extraordinary concern for life after death (vulture or scorpion) on the other.

The third period of Egyptian culture was followed by the fourth period, the personality culture of the Greeks and Romans. It was under the influence of that zodiac force, which in the souls of the people of earlier times evoked the image of a ram looking back on itself. This new influence found expression at that time in the myth of the Golden Fleece or hide of a ram, in the preference of the Hebrews and Arabs for eating lamb, in the designation of Christ as the Lamb of God or his representation as the good shepherd leading the flock of lambs, i.e. the humanity of the Age of Aries.

By the interaction of the zodiac force of Aries with the opposite one of Libra, private law was developed and, in general, a further development of law in Roman culture. Both played a significant role in the upbringing and education of people at that time, as Rudolf Steiner reports:

"This general human education changes from period to period. Nowadays, we are already unhappy, when our children reach the age of ten and can't yet calculate certain things, aren't we? The Romans were not yet at all. But they were unhappy if such a boy did not yet know the Laws of the Twelve Tables, while we again take less care that our children know the provisions of the law." [38]

Even today, we can find sculptures of Lady Justice with the scales in many courthouses as an after-effect of that time. Christ was also depicted as the great judge of the world, who ensures a balance between the justified claims of the individual and those of the general public, between the individual and the whole of humanity. Unfortunately, this influence from the fourth cultural period is nowadays misused by some traditional religious communities to judge people and label them as sinners if they do not fit into a rigid grid of outdated ways of thinking. In some cases, they are even denied their right to exist and, in extreme cases, their murder is justified. This shows how originally justifiable influences of the zodiac forces are always subject to the danger of being reinterpreted by people in the course of the centuries and, in the context of rigid formation of traditions, actually turned into the opposite.

In 1413, a new cultural development began again. Its centre is in the part of Europe north of the Pyrenees and the Alps. Many centuries before, this culture was being prepared under the influence of the ancient Roman World and Christianity as its state religion. The influence of the zodiac force of Aries was followed now, in the fifth cultural period, by that of Pisces. The clear gaze of Aries was still strongly directed towards external sensory perception. The zodiac

---

[38] GA 194 "Die Sendung Michaels – Die Offenbarung der eigentlichen Geheimnisse des Menschenwesens" (Mission of Michael – The revelation of the actual secrets of the human being), Dornach, lecture of 6 December 1919.

force of the water sign Pisces, on the other hand, has more of an effect on the inner soul life by stimulating the development of images in the soul. Pisces is a visionary sign and under its influence people will very gradually regain an inner vision of images, an Imaginative consciousness, but this time while fully maintaining the waking day and "I"-consciousness. The old passive, dream-like clairvoyance will be replaced by a new kind of active, conscious clairvoyance. This development is still in its infancy. It first expresses itself in an intensification of the life of imagination. The ideas and inner images of the soul are initially as fleeting and difficult to grasp as fish in water. Many people are not yet really aware of the development of this new inner level of experience. Others are already aware of it, but feel worried, insecure or even frightened because they do not understand what is stirring in their souls. Through anthroposophy we know that the ability of a new type of consciousness, Imagination, is slowly creeping in and gradually developing here, which will enable us to access the next higher level of existence, the etheric-astral world.

Today, the prevailing opinion is that in antiquity people thought up their ideas about the world with a great deal of fantasy, and that only we modern people have progressed to a realistic way of looking at things. Basically, however, the opposite is the case. The people of antiquity looked very closely at the outside world. They did not think about things as abstractly and intellectually as we do today. Rather, they had the feeling that the external impressions evoked in their consciousness, in addition to the sensory image, also the associated thoughts and concepts. What else they added were the traditions from preceding cultures. Their mythologies, however, were not based on any arbitrary fantasies, but on real, albeit inner, perceptions of earlier humanity. In the case of the Greeks, however, these were already very over-formed and drawn down into the human, because they lived very much in outer perception and physical existence in the physical world. In our present, fifth culture, on the other hand, we have become very superficial as far as sense perception is concerned. Under the zodiac force influence of the Mercury sign of Virgo, which is opposite Pisces, we are developing the intellect and independent thinking rather than perception. But this is still in the early stages of its development.

"One only looks at the world in a certain short-sighted way today. Today's peasant t h i n k s more than the Greek philosopher thought. In contrast, the faculty of perception was much more developed in those days. Man was connected to the whole of nature. Then, perception was the same as what the imagination is now to us. Today, perception is no longer learned at all, only by those who undergo training. It is quite possible for a person to get far in what he learns in the laboratory, and yet be very inexperienced outside, not being able to distinguish wheat from rye. So we can say that people today have a lot of imagination, but at that time they were trained in perception. Therefore we can distinguish two epochs: an epoch of perceptions and one of imaginations." [39]

The modern, scientifically trained and accordingly thinking human being lives in imaginations of atoms and elementary particles and such fantastic ideas as the alleged big bang of the universe, which after a primeval explosion miraculously and all by itself formed the great world order, the cosmos (Greek word for order). For entertainment, we indulge in feature films and young people immerse themselves in computer games that make them believe in fantastic illusory worlds. Some people would rather live in such worlds than in the mundane, real world, and the entertainment

---

[39] GA 143 "Erfahrungen des Übersinnlichen – Die drei Wege der Seele zu Christus" (Experiences of the supersensible – The three ways of the soul to Christ), Winterthur (Switzerland), lecture of 14 January 1912.

industry is working diligently to fulfil this wish ever better. Instead of developing the power of imagination in one's own soul together with the power of thought to the next higher form of consciousness, the Imagination, many prefer to be passively exposed to a constant stream from outside. The number of those who use chemical drugs to escape reality and the tasks of earthly life is growing worryingly. But if the new forces that want to emerge in humanity are not actively grasped and developed in a healthy way, they just show themselves in a chaotic way. Accordingly, the demand for psychological and medicinal help is increasing.

Anthroposophy is a great support to us here by enlightening us about the changes of the soul in the course of the different ages and cultures. What you understand is less frightening. In addition, purely spiritual thought contents and ideas have an ordering, calming, strengthening and healthy effect on the human soul life. Gradually, more and more people will benefit from this. The further development in the course of our fifth culture will really push them in this direction.

However, all development always takes place in many small steps. Thus, the individual ages or cultures are divided into periods, each of which has its own special tasks and accordingly develops its own unique forms of expression. For example, historians divide the period of the last millennium into the cultural sections Romanesque, Gothic, Renaissance, Baroque and the Modern or Industrial Age. If one first carefully absorbs the information available about these periods in a purely mental way, with the intellectual zodiac force of Virgo, which is active in today's Pisces Age as well, and then allows everything to work in the soul in a meditative mood, the zodiac force of Pisces, which leads to future vision of images, transforms the thoughts into pictorial ideas. For many people these may still be pale and inconspicuous, so that they may not even notice this process. But those who are already somewhat practised in placing images in their consciousness during meditation can already actively deal with this force, however initially it may be developed. You can then let the individual cultural periods pass before your inner eye in the form of inner images, as if on a timeline. This reveals to the observer that the qualities of the individual periods of an age also follow the order of the zodiac forces quite faithfully and that it is precisely these that give the periods their respective character.

They can be determined more precisely in terms of time if one simply divides the average duration of the ages of 2,160 years given by Rudolf Steiner by 12. This results in 12 periods of 180 years each. If we now want to take a closer look at the cultural development of Europe in the course of the last thousand years, we only need to subtract 180 years twice from the year 1413, the beginning of the Age of Pisces, to get close to the year 1000 AD. The result is the year 1053, which corresponds to the beginning of the second last period of the Age of Aries.

People in Europe lived very close to nature at that time. The earth with its natural gifts was regarded as a gift from God, and European culture was essentially shaped by the Christian monasteries. These all had their monastery gardens, not only for the cultivation of food, but also for the care and cultivation of medicinal plants. The strong orientation towards nature and the earth found its expression even in the architecture and art of the Romanesque, which, after a preparatory phase, found a remarkable flowering in the High Romanesque period from about the first century of the new millennium. At that time, large churches with several naves were built. For this, the problem of a safe and stable vaulting had to be solved. This was achieved in the form of compact, massive, earth-bound basilicas with their typical round arches. Everything was built according to the principle of safety and stability. At the same time, however, the aim was also to express beauty.

Not only horticulture was practised, but also social life was now to be made more cultivated and beautiful. The ideal of the courageous and energetic, but somewhat rough-around-the-edges knight of the preceding cultural period was to be replaced by that of a knight who knew how to behave at court, a knight who behaved "courtly" or, as we still say today, "courteously". Along with this, the previously impetuous wooing of the knights, which was primarily aimed at "conquest", changed into a more cultivated form, the so-called "Minne" or courtly (courteous) love. It became a central theme in literature, song and painting from the second half of the 11th century.

These are all qualities that correspond in the zodiac to the force of the earth and Venus sign Taurus. This heyday of the Romanesque lasted until the 13th century. Its name indicates that it was still part of the Roman culture of the fourth age. It was its penultimate period and we can call the 180-year period from 1053 to 1233 the Taurus period of the Aries age, which was gradually coming to an end:

1053 – 1233 AD	High Romanesque Taurus period (penultimate period) of the Aries Age

Opposite Taurus in the zodiac is Scorpio. Similar to the ancient Egyptians, the cult of the dead therefore played a special role in the Taurus period of the Aries Age. The most important component of every Romanesque cathedral was the crypt, the tomb of a saint or at least a high dignitary. The main altar was erected directly above it. The altar itself had and still has the shape of a stone coffin in Christian churches. This is not only an imitation of the situation of the early Christians in Rome who, because of their persecution, had to retreat to underground catacombs for their services, where they buried their dead, but also has to do with the fact that the zodiac forces of Taurus and Scorpio are associated with the two principles of life and death. It is true that Christ did not pass through death and rise to life again in the Age of Taurus of the ancient Egyptians, but only in the Age of Aries. However, the Mystery of Golgotha occurred in the fifth period of this age, in which the zodiac force of Scorpio (death) and its opposite sign Taurus (life as an etheric power) prevailed.

This statement may at first seem surprising, for Scorpio, as we know, is not the fifth but the eighth sign of the zodiac. But, it must be taken into account here that the first 180-year period of an age is always subject to the zodiac force of Cancer. Similar to the way a human being is born out of the womb of a mother, a new culture always begins under the influence of the mother sign Cancer.[40] The first four sections of the Age of Aries, each lasting 180 years, were therefore under the rule of the zodiac forces of Cancer, Leo, Virgo and Libra. With the expiration of these four periods, the first third of the Age of Aries came to an end. Now followed the fifth or Scorpio period from 27 BC to 153 AD. After the first third had expired as well, exactly 60 years after its beginning, the Mystery of Golgotha took place in 33 AD.

Here we can already see that the order of the zodiac forces in the small cultural periods lasting 180 years does not correspond to the backward order as we find it in the ages as a result of the precession of the vernal point, but to the normal order in the course of the year.

---

[40] The biological evolution of living beings on earth also begins with the zodiacal force Cancer (GA 300c "Conferences with the Teachers of the Waldorf School", Volume 3, Conference of 12.07.1923, 8 p.m.).

The further considerations on the individual phases of the cultural development of Europe will confirm this.

1233 – 1413 AD High Gothic
Gemini period (ultimate period) of the Aries Age

From the compact earthiness of the Romanesque period, there was now a transition to a construction method that reached for the sky. Slender, towering cathedrals rose from the earth high into the air and the endeavour to let as much light as possible stream into the interior through gigantic glass windows pushed back the load-bearing capacity of the masonry to such an extent that the walls of the churches had to be supported on the outside by a finely thought-out system of buttresses. The effectiveness of a new and quite different zodiac force is clearly expressed here. It is that of the air and Mercury sign Gemini. Just as Mercury, the messenger of the gods, rose into the air, so one wanted the prayers of the people in the churches to flow up to God in heaven, in harmony with the cathedral surrounding them.

Mercury is also the god of trade. Accordingly, a new middle class grew up in society, alongside the nobility and the clergy, which achieved prosperity through trade. The guilds gained increasing economic and political power in the towns. From the preliminary form of the so-called "Kaufmannshanse" (merchants' Hanseatic League), the so-called "Städtehanse" (towns' Hanseatic League) finally developed as a supra-regional trade association in the period from 1233 to 1413. Many trade routes were fortified and the European trade network was further expanded during this time.

The zodiacal image of Gemini is opposed to that of Sagittarius. As already in the primeval Persian culture, the Twins, who were originally seen supersensorily as a human couple, represent the earthly human being. Sagittarius was seen as a centaur, as an Imaginative expression of the higher, celestial human being, who rises with the higher parts of his soul out of the lower ones and directs his gaze and the arrow of his bow, i.e. the power of his soul, towards heaven. The Gothic cathedral united this polarity by being viewed, with its extraordinary abundance of light and splendour, as an earthly image of the heavenly Jerusalem, the abode of the community of saints who have already completed the development into the higher, heavenly human.

In this field of tension between the earthly and heavenly worlds, it was possible to bring human thinking a great step forward in its development under the influence of the Mercury sign Gemini. The techniques of logic and dialectics, which had been preserved from the time of ancient Greece, were used. They had been established by the philosopher and natural scientist Aristotle[41] in the Virgo period of the Age of Aries (387 - 207 BC)[42], which was also under the influence of Mercury. Now, under the decisive leadership of Dominican monks, above all the Italian Thomas Aquinas[43], the Aristotelian method of thought and concept formation could be further developed into High Scholasticism. But it restricted human thought to the investigation of the earthly world of the senses. Everything that was attributed to the heavenly world was left to faith. In this field, some contemporaries sought to attain higher knowledge and insight by way of feeling. Under the Jupiter influence of the zodiac force of Sagittarius, Christian mysticism flourished in the Middle Ages, the best-known

---

41 Aristotle (384 – 322 BC)

42 see p. 62

43 Thomas Aquinas (1225 – 1274 AD)

representatives of which were Meister Eckhart and Heinrich Suso[44]. However, their spiritual experiences were only possible through the suppression of the "I"-consciousness.

With the year 1413, not only began a new cultural period of one hundred and eighty years, but also a completely new cultural epoch lasting 2,160 years, the so-called fifth culture or the Age of Pisces. Its task is to develop in people above all the ability to experience inner pictorial ideas or imaginations under the influence of the zodiacal force of the Jupiter sign Pisces, and the ability of rational thinking, the intellect, under the zodiacal force of the Mercury sign Virgo. Through the interaction of both forces, the life of the imagination can develop while retaining full awake "I"-consciousness, not in a dreamlike manner as was the case in earlier millennia with the soul images of the ancient clairvoyance.

But every new development first begins with a kind of repetition of what has gone before, a looking back on the previous age, which leads to a temporary revival of the same. Accordingly, even our modern historians refer to the first phase of the new age with the French term "Renaissance", i.e. rebirth. It took place, like every birth, under the influence of the mother sign Cancer, which gave the first 180-year period of the new age its very special character.

1413 – 1593 AD	Renaissance
	Cancer period (first period) of the Pisces Age

[44] The German Mystics Eckhart von Hochheim, commonly known als Meister Eckhart (c. 1260 – c. 1328), and Henry Suso or Seuse (1295 - 1366)

After a preparatory development first in Italy, from the beginning of the 15th century a strong tendency to look back to the Greco-Roman antiquity that had come to an end, also spread among the European peoples north of the Alps. Latin was still the international language of communication and teaching anyway. But now the ancient Greek language was also taught at the universities and many ancient manuscripts were rediscovered.

Greco-Roman culture had been shaped by the influence of the zodiac forces of Aries and Libra. They had led people out of their tribal and group consciousness to the first stage of an individual consciousness, which was manifested in Roman private law, which was contrasted with the law of the general public, of the state. At the beginning of the fifth culture, however, there was no mere repetition of the fourth culture, but its main theme, the polarity of "individual and generality" or "man and world", was now given a special character by the zodiac force of Cancer and its antipole Capricorn, which dominated the first period of the new age, the 15th and 16th centuries.

The watery sign Cancer is a sign of inwardness. Like Pisces, it directs a person's attention to the inner world of his soul. The earthy sign Capricorn, on the other hand, is a sign of the outer world. Under the influence of Cancer, individual consciousness, which had already been established in antiquity, now clearly increased in strength. The dignity of the human being became an issue. Humanism came to blossom. Mankind felt itself to be the crown of creation. The nobility and the bourgeoisie wanted to see themselves depicted in portraits and sculptures with individual features, dignified, in a moving pose and with strong emotional expression. One should see what they felt. This aspiration finally culminated in Mannerism, the style of art of the Late Renaissance with its exaggerated poses.

At the same time, the zodiac force of Capricorn directed people's interest towards the physical outside world, which they now sought to explore more closely in purely intellectual terms. Humanism was confronted with naturalism. With ambition and perseverance, the naturalists of that time dared to tackle great tasks. And since the water sign Cancer was the primary and predominant sign at that time, people longed to go out into the vastness of the unknown and dangerous oceans, whereas before they had mainly navigated the rivers, coastal waters and inland seas. The Renaissance thus became the age of the seafarers to a special degree. The aim was to reach the limits of the physical world, indeed to circumnavigate and encompass the whole world. Following this impulse, the Italian Christopher Columbus[45], who was in the service of Spain, gave the impetus for an enormous transformation of the image of the earth at that time by rediscovering America.

In a similarly revolutionary manner, the canon Nicolaus Copernicus[46], who lived in the East Prussian region of Warmia[47], penetrated into the regions of extraterrestrial nature by observing and calculating. He completely rearranged the world of the planets according to a purely physical view by replacing the geocentric world view of Ptolemy, which was still predominant at the time, with a heliocentric one. It is no coincidence that he chose the ambiguous title "De Revolutionibus" for his work, using the Latin word "revolutio", which means both "revolution" and "upheaval".

Even the religious life of European humanity now underwent a massive change. 1,500 years after the founding of Christianity, it had

---

[45] Christopher Columbus (1451 - 1506)

[46] Nicolaus Copernicus (1473 - 1543)

[47] Warmia (German: Ermland) was a German settlement and autonomous prince-bishopric in East Prussia that had voluntarily placed itself under the protective sovereignty of the Polish king in 1466.

become extremely externalized under the leadership of the Catholic Church. Ecclesiastical offices had long since become offices of power and were occupied mainly by nobles whose material interests outweighed their meagre spiritual interests many times over. In the practical life of faith, too, people relied primarily on material things, such as the physical relics of deceased saints. The worship of relics and the trade in relics took on almost absurd features. The Church itself provoked the superstition among the people that one could buy one's afterlife salvation through material gifts. These blatant conditions caused a strong longing for a return to genuine Christian values and the inwardness of faith, which ultimately led to the Reformation under the leadership of the German Augustinian monk Martin Luther[48].

Unfortunately, the ability to see the spiritual world, which people still had in ancient times, had long since disappeared. Whether they wanted to or not, they could only rely on external things. The written traditions of Christianity, summarised in the Bible, served this purpose. The two thousand years and even older formulations in it were translated and interpreted from the perspective of the Middle Ages. What was believed to be understood was called the Christian faith, which, under the motto "Sola fide, sola scriptura, solus Christus, sola gratia" (by faith alone, Scripture alone, Christ alone, by grace alone), opposed the outward pomp, money and power of the papal church.

Thus the new Piscean Age began with three powerful drumbeats in the course of its first cultural period: a substantial expansion of the concept of the earthly world through the rediscovery of America, a fundamental transformation of the extraterrestrial world as well

48 Martin Luther (1483 – 1546)

through the founding of the heliocentric world view, and a radical division of the European Christianity into Protestants and Catholics.

Towards the end of the 16th century, the second one 180-year period of the new age began: the Baroque. The predominant zodiac force was now that of Leo.

1593 – 1773 AD	Baroque Leo period (second period) of the Pisces Age

The emphasis on human dignity that had been strengthened in the preceding Renaissance period grew into a pride in one's own personality. Those who could afford it presented themselves more and more splendidly. In order to elevate themselves, people wore wigs which, in the case of women, reached upwards and, in the case of men, were modelled on a lion's mane. Under the influence of the sun and fire sign Leo, people shunned water and even considered it unhealthy to wash. Instead, they preferred to powder and perfume themselves. Lion-like leaders were very respected, so that the monarchy flourished to the highest degree up to French absolutism. The French ruler Louis XIV[49] is certainly the most extreme example of this new attitude. He saw himself as the sun, outshining and dominating the world with his brilliance. Eventually he was given the epithet "Sun King".

The people still lived in the medieval view that monarchy was a order willed by god. But increasingly, the "I"-consciousness and the wish for recognition of the dignity of each individual human being was also gaining strength among them. As a result of the very passionate and fiery emotional life of the time, there were fierce

---

49 Louis XIV (1638 – 1715), French king

conflicts between representatives of different world views and opinions. This eventually led to the devastating Thirty Years' War (1618 - 1648) between Protestants and Catholics, which destroyed countless lives and completely devastated vast areas of Europe.

In natural science, the astronomers Johannes Kepler[50] and Galileo Galilei[51] consolidated the new world view which placed the sun at the centre of the planetary system , founded by Copernicus right at the beginning of the Age of Pisces, but still rejected by the Church. In his law of gravitation, Isaac Newton[52] described the centripetal force of gravitation emanating from a centre and pulling towards this centre, completely in line with the idea of centralism that prevailed at the time.

But, under the influence of the zodiac force of Aquarius, the scientific way of thinking and humanism of the preceding Renaissance period gave rise to the intellectual school of the Enlightenment as an antithesis to the zodiac force of Leo. Free thinking based on reason, which was subject neither to the dictates of a monarch nor to those of the church, led to the view that freedom, equality and fraternity were natural rights of all people and should be the basic principles of their coexistence. This counter-movement to monarchism brought about parliamentarism in England and, in 1776, the first democracy on the North American continent, where a new form of state could be established unhindered by the entrenched, old, grown structures of Europe. In France, the further development culminated in 1789 in the revolution of the people and

---

50 Johannes Kepler (1571 – 1630), German astronomer, astrologer and mathematician

51 Galileo Galilei (1564 – 1642), Italian astronomer, physicist, mathematician and engineer

52 Isaac Newton (1643 – 1727), English astronomer, physicist and mathematician

the bloody overthrow of the kingship. With such drumbeats, the Leo period came to an end and the following Virgo period was ushered in.

1773 – 1953 AD	Neoclassicism / industrial era Virgo period (third period) of the Pisces Age

The pomp of the Baroque was replaced by the simpler, more stricter forms of Neoclassicism. Further development took place under the influence of the zodiac forces of Virgo and its counterpart Pisces. Thus, the same forces that shaped the entire 2,160-year Piscean Age (1413 - 3573 A.D.) were at work in this new cultural period, which again lasted 180 years. Accordingly, from the transition to the 19th century, there was a further development of the pictorial power of imagination (Pisces) on the one hand and the analysing power of reason (Virgo) on the other, with the latter predominating, since it was now the Virgo period of the Age of Pisces. Under its Mercury influence, the medieval thinking technique of scholasticism was transformed into the critical scientific way of thinking.

The first results found a typical expression as early as the end of the 18th century in the works "Critique of Pure Reason" (1781), "Critique of Practical Reason" (1788) and "Critique of Judgement" (1790) by the German philosopher Immanuel Kant[53] and made theoretical materialism the predominant view of the world and man in the 19th century. The power of understanding could be best developed, in the beginning, on the gross, external world of sense. In this early stage, it did not yet reach the insight that human cognition is by no means limited to the outer world of the senses, but that it is

53 Immanuel Kant (1724 – 1804)

destined to penetrate into psycho-spiritual, and thus supersensible levels of existence in the future. Natural science will find its extension in a science of soul and spirit, and thus arrive at a more complete understanding of the world and the human being. Rudolf Steiner's[54] anthroposophy opened up a first viable path to this end. At first, however, people's gaze remained focused on material nature.

New discoveries and technical inventions made it possible to build ever more complicated machines. Thus industry became the determining factor of the new cultural period. A class of the population toiling in the factories emerged, the proletariat. Its poverty ultimately led to the workers' movement. The serving and working people fought for their rights. The power of the monarchs and aristocrats was increasingly curtailed. This development, which led to the formation of parliaments elected by the people or even a democracy, the rule of the people, had, like all new developments, already taken its beginning in the second half of the preceding cultural period. Now it clearly came to the fore.

Besides the principles of work and service, the zodiac force of Virgo also includes the principles of health and hygiene. Accordingly, from the 19th century onwards, ever greater importance was attached to cleanliness and disinfection in medicine. In this way, great progress was made in surgical procedures and wound treatment.

The dry intellectual thinking under the influence of the earth and Mercury sign Virgo was polar-contrasted with the further development of the imaginative life under the Jupiter sign Pisces. The latter was expressed in the new cultural trend of Romanticism. It attached more importance to feeling and fantasy than to abstract thinking. The Romantics turned with great interest to the medieval world of

---

54 Rudolf Steiner (1861 – 1925), Austrian, founder of Anthroposophy

legends and fairy tales, to mysticism and the supernatural. The Brothers Grimm[55], for example, were representatives of this movement and published their collection of German folk tales. One of the most outstanding German Romantics was certainly Georg von Hardenberg[56], who wrote under the pseudonym Novalis. His contemporary Goethe[57] met Novalis in person and was, like Novalis, both a scientist (Virgo) and a poet (Pisces), turned above all against the tendency towards the fantastic, which was so popular among the Romantics. Here, he complained especially about the lack of reference to reality and truthfulness. Nevertheless, Goethe in particular already developed a very lively imaginative life, by means of which he sought to overcome the "limits of cognition" set by Kant and to penetrate into the supersensible spiritual that underlies all that is physical. Rudolf Steiner reports in his autobiography "Mein Lebensgang" (The story of my life) about a significant conversation between Goethe and his fellow poet Schiller[58] in this respect:

"And Goethe drew his «archetypal plant» before Schiller's eyes with a few strokes. It depicted the plant as a whole through a sensible-supersensible form, out of which leaf, blossom, etc. shape themselves, reproducing the whole in detail. Schiller, because of his Kantian point of view, which he had not yet overcome, could only see an «idea» in this «whole», which human reason forms by observing the details. Goethe did not want to accept this. He «saw» spiritually the whole as he saw sensorily the detail. And he did not admit any difference in principle between spiritual and sense perception, but only a transition from one to the other. It was clear to him that both claim to stand in the experiential reality. But Schiller could not get away from maintaining

---

55 Jacob Grimm (1785–1863) and Wilhelm Grimm (1786–1859)

56 Georg Philipp Friedrich von Hardenberg (1772 – 1801)

57 Johann Wolfgang von Goethe (1749 - 1832)

58 Friedrich Schiller (1759 – 1805)

that the archetypal plant was not an experience but an idea. Then Goethe replied, out of his own way of thinking, that he would just see his ideas with his eyes." [59]

This shows how the early beginnings of a new clairvoyance appeared in Goethe as early as the 19th century, while retaining the full, alert I-consciousness and the power of reason, as a result of the interaction of the zodiac forces of Pisces and Virgo, which are decisive for the Age of Pisces or the fifth culture.

The analytical, neatly separating thinking of Virgo caused in the period from 1773 to 1953, unfortunately, a strong wish among the peoples for strict, clear demarcation from the neighbouring peoples. Narrow-hearted nationalism and thinking of dominance spread. Every nation considered itself the greatest. The life of imagination, which grew chaotically for many under the influence of Pisces, stirred up emotions and, together with ice-cold intellectual thinking, led to unrealistic ideologies, which found their most terrible expression in the Holocaust. National egoism and an enormous tendency towards untruthfulness among leading politicians of the time ultimately led to the catastrophe of the two great world wars. [60] The third cultural period of the Piscean Age ended with these loud bangs. Europe was once again in ruins.

With the reconstruction, a new, more peaceful time began. Under the influence of the zodiac force of Libra, the realisation grew that

---

[59] GA 28, Rudolf Steiner, "Mein Lebensgang" (The story of my life), chapter V.

[60] See e.g. Rudolf Steiner's lectures in GA 173 and 174 "Zeitgeschichtliche Betrachtungen – Das Karma der Unwahrhaftigkeit" (Historical considerations – The karma of untruthfulness), GA 174a "Mitteleuropa zwischen Ost und West" (Central Europe between East and West) and GA 174b "Die geistigen Hintergründe des 1. Weltkrieges" (The spiritual background of World War I).

peaceful coexistence of peoples is only possible if one focuses on what is communal and not on what divides.

1953 – 2133 AD	Peace movement, environmental protection and equal rights for all people Libra period (fourth period) of the Pisces Age

The main interest of the post-war period was initially to overcome the material hardship that existed everywhere. To do this, the economy had to get going again. But this was now to be done in international cooperation. The European Coal and Steel Community (ECSC) was founded as early as 1951. In 1957, the European Economic Community (EEC) and the European Atomic Energy Community (EURATOM) were added by the Treaty of Rome. In 1993, the EEC was transformed into the European Union (EU) by the Treaty of Maastricht, incorporating further common areas of responsibility, which was to promote many other common concerns beyond the economy and become a community of values. Since the beginning of this international development, differences of opinion between the peoples of the West are no longer fought out martially with weapons, but discussions are held to find common solutions in a peaceful manner. And since the zodiac force of Libra strives to form a world community, numerous global organisations were founded in the post-war period, as early as 1945 the United Nations (UN) and its meanwhile 17 sub-organisations, the best known of which are UNESCO (Education, Science and Culture), WHO (World Health Organisation) and UNICEF (Children's Fund). Many other global organisations were added later.

But, it must be taken into account that the special characteristics of each cultural period still extend far into the following one and only gradually fade out there. Accordingly, we still find the after-effects of dividing nationalism in today's Libra period. A typical example of this is the construction of the Berlin Wall in 1961 as a border fortification between the Federal Republic of Germany (FRG) and the German Democratic Republic (GDR). Even at the beginning of the third millennium, in which we now live, nationalist aspirations flare up every now and then due to backward-looking attitudes. Development proceeds in waves, according to the principle of two steps forward and one step back.

In the previous Virgo period, mankind was mainly interested in construction: industrial production. The burning of fossil fuels produced vast quantities of exhaust gases, which were simply emitted into the atmosphere. All products that had become unusable after use were deposited in landfills, polluting the soil and groundwater. In an effort to improve hygiene, all impure substances and sewage were channelled through underground sewers into rivers, lakes and seas. Mankind had not yet matured to the realisation that life on earth can only be sustained if all construction is matched by corresponding dismantling. This realisation is only now beginning to take hold under the influence of the autumn sign Libra. Nature knows very well how to dismantle everything it builds up and use it for new construction. Mankind, on the other hand, has yet to learn recycling.

The zodiac sign Libra, like the zodiac sign Taurus, is a Venus sign and therefore very close to the subject of love as well as to the life forces and processes of nature. Accordingly, "Make love not war" became the new motto and the subject of sexuality, which had been virtually taboo in the previous prudish Virgo period, now came to the fore. Sexual education became part of the school curriculum. Parallel to this, interest in organic and biodynamic agriculture and animal husbandry grew. Nature and climate protection have now become the dominant issues of our time. The new goal is the conversion of the energy industry to environmentally friendly, renewable energies such as solar energy, wind power and hydropower.

Although this will not stop climate change, it will at least slow down the rise in temperature somewhat and dampen its dramatic effects on humanity. For just as every winter is followed by a summer in the course of a solar year, the climate oscillates between ice ages and hot ages in the course of a Great World Year. In the Platonic Year, however, a reversed regularity prevails in comparison to the solar year. In the latter, the sun moves forward through the zodiac. During a Platonic Year, the rising point of the Sun at the beginning of spring moves backwards through the zodiac. Warm and cold periods are also distributed in exactly the opposite way. In the solar year, the hottest weeks, on average, fall in July/August, when the Sun moves through the power areas of Cancer and Leo. In the Platonic year, exactly the opposite is the case. When the vernal point moves backwards through Leo and Cancer, we have an ice age on Earth. On the other hand, when the vernal point passes through Aquarius and Capricorn, the two signs under whose influence we experience the coldest weeks in the solar year, there is a hot age on Earth. During the present Piscean Age, which immediately precedes the Aquarian Age, a clear rise in temperature is therefore already noticeable. In addition to this process there is man-made global warming due to the enormous amount of greenhouse gases emitted from the burning of fossil fuels. They accelerate and intensify the global warming process. Humanity is therefore called upon to take countermeasures here.

Opposite Libra in the zodiac is Aries, the sign of the single person, of personality and individuality. Accordingly, human rights are another major issue of our time. More and more people are realising that being human means diversity and that one can only do justice to the different varieties of being human if they are allowed to stand side by side on an equal footing, all complementing each other to form a great unity. The zodiacal forces of Libra and Aries, in their interaction, are therefore the driving forces behind the striving for equal rights for all people, regardless of ethnic origin, religion or world view, biological sex or gender self-identification, sexual orientation or any disability.

A highly interesting phenomenon in this context is trans-sexualism, which has recently come into the public eye, and which should more correctly be called transidentism. It proves that the soul of a human being is something completely independent and by no means only a product of the body, as natural science claims. A completely normally developed male body with an intact testosterone balance cannot possibly produce a human soul that feels itself as female, just as a normally developed female body with an intact oestrogen balance cannot possibly produce a human soul that feels itself as male. If the soul were a product of the body, its sensation would have to correspond to the nature of its producer. Religious communities that teach the immortality of the human soul should therefore gratefully address the phenomenon of trans-identism instead of rejecting it as “unnatural”.

What is unnatural, on the other hand, is the brick-headed view that only the two manifestations “heterosexual masculine man” and “heterosexual feminine woman” correspond to God’s will. Such a view profoundly disregards the diversity in God’s creation and demonstrates a blatant lack of knowledge of the world and humanity. It is just as foolish as if one were to divide the year only into the polarity of summer and winter and judge the transitional periods of spring and autumn as unnatural. On the other hand, a closer look reveals that the year manifests itself in twelve different forms, the twelve months, corresponding to the twelve forces of the zodiac.

They underlie the whole of creation and thus also human beings as a structuring principle.

It is to be hoped that the conception of man of past centuries, based on a coarse and narrow-minded way of thinking, will finally be overcome in the further course of the present Libra period, which will last until the year 2133, and replaced by a conception of man that is worthy of him. After all, Aries and Libra are precisely those zodiac forces that also ruled the age in which the Christian impulse entered humanity with the aim of bringing into balance human individuality in all its diversity on the one hand and the concerns of the whole of humanity as a unity on the other. Only in this way will a peaceful, understanding and loving coexistence of all people on earth be possible.

In summary, it can be said that the considerations made here on human history not only prove that each age develops its very characteristic appearance under the influence of a special zodiac force, but moreover that the ages in their succession also correspond exactly to the sequence of the twelve zodiac forces, albeit in reverse order due to the retrograde precession of the vernal point.

Within each age, twelve cultural periods can be observed, which also receive their character from the zodiac forces, although here the sequence corresponds to the "normal" order of the zodiac signs, as we know it from the movement of the sun in the course of the year.

Both together open up a deep insight into cultural history and basically make it understandable why many things have developed in the way they have, and even why they happened in that order.

List of Figures

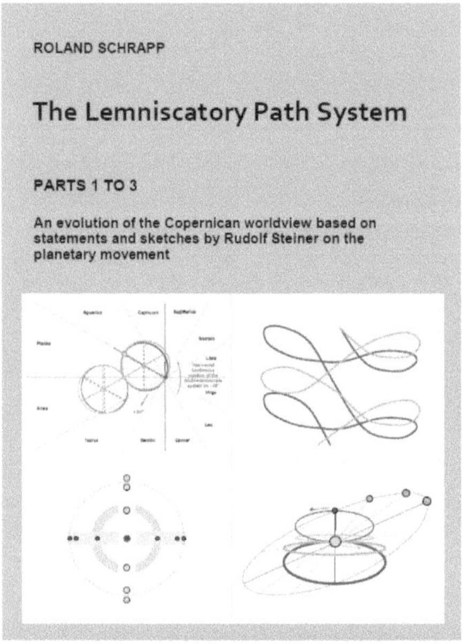

Roland Schrapp

Publisher:
BoD – Books on Demand,
Norderstedt (Germany)

Paperback: 198 pages
Large format (DIN A4)
253 mostly coloured illustrations

ISBN-10: 3752604034
ISBN-13: 978-3752604030

An evolution of the Copernican worldview based on statements and sketches by Rudolf Steiner on the planetary movement.

For the first time in almost a hundred years, Rudolf Steiner's statements and sketches on the subject of the "Lemniscatory paths of the planets", distributed over several lecture cycles, have been brought into a larger context and examined for the consequences of this. Steiner's suggestions for a new consideration of the planetary movement were taken up and tried to develop them further in the given sense. The work "The Lemniscatory Path System" arose from this. The treatise comprises 192 pages with 253 mostly coloured illustrations.

Roland Schrapp

Publisher:
BoD – Books on Demand,
Norderstedt (Germany)

Large format (DIN A4)
270 pages, 27 illustrations

Paperback (adhesive binding):
ISBN-10: 3752690100
ISBN-13: 978-3752690101

Hardcover (thread binding):
ISBN-10: 3752602902
ISBN-13: 978-3752602906

This book takes a completely new look at the Anthroposophical Soul Calendar. It is about the deeper meaning of the fifty-two weekly verses, which has remained essentially unexplored in the last hundred years since the first edition by Rudolf Steiner. A dense veil of Isis was spread over them, of which is well known that no mortal person can lift it. Only the immortal, psycho-spiritual human being, who knows himself at home in the extrasensory, higher worlds, is capable of doing this. Only to him the weekly verses reveal themselves as a travel guide through these worlds and lift him up to ever higher spiritual-cosmic realms until he reaches the experience of God, from where he gradually descends again into a new life on Earth, enriched in spirit and fertilized in his soul. If the reader embarks on this journey, the spiritual archetype of the Soul Calendar is ultimately unveiled to him and he achieves an extended understanding of Man and Christ. By many quotations from Rudolf Steiner's lectures and books, the author virtually lets Steiner himself elucidate the breathtaking depths of his mysterious weekly verses.